趣懂Java旋律
击破36个难点

耿祥义 张跃平◎主编

清华大学出版社
北京

内容简介

本书按照场景故事、场景故事的目的、程序运行效果与视频讲解以及阅读源代码的模式，通过有趣的故事或场景体现学习 Java 的主旋律，帮助读者理解 Java 程序设计中的关键点和难点。

本书选取的场景故事言简意赅、诙谐幽默，充分体现相应的编程概念、思想和方法。读者通过 36 个有趣或熟悉的场景故事和 430 分钟的视频讲解，能够掌握 Java 程序设计的核心概念、面向对象的思想和重要的设计模式。

本书可作为计算机相关专业本科生 Java 课程的参考教材，也可供软件设计开发人员参考使用。

图书在版编目（CIP）数据

趣懂Java旋律，击破36个难点 / 耿祥义，张跃平主编. — 北京：清华大学出版社，2021.2
ISBN 978-7-302-57048-6

Ⅰ. ①趣…　Ⅱ. ①耿… ②张…　Ⅲ. ①JAVA语言—程序设计—高等学校—教材
Ⅳ. ①TP312.8

中国版本图书馆CIP数据核字（2020）第238169号

责任编辑： 陈景辉
封面设计： 刘　键
插画设计： 刘　昉
责任校对： 徐俊伟
责任印制： 杨　艳

出版发行： 清华大学出版社
网　　址：http://www.tup.com.cn，http://www.wqbook.com
地　　址：北京清华大学学研大厦 A 座　　邮　　编：100084
社 总 机：010-62770175　　邮　　购：010-83470235
投稿与读者服务：010-62776969，c-service@tup.tsinghua.edu.cn
质量反馈：010-62772015，zhiliang@tup.tsinghua.edu.cn
课 件 下 载：http://www.tup.com.cn,010-83470236
印 装 者： 大厂回族自治县彩虹印刷有限公司
经　　销： 全国新华书店
开　　本： 180mm × 240mm　　**印　　张：** 15.25　　**字　　数：** 281千字
版　　次： 2021年4月第1版　　**印　　次：** 2021年4月第1次印刷
印　　数： 1~2000
定　　价： 59.90元

产品编号：088948-01

PREFACE 前言

在学习Java的过程中，理解和掌握重要、关键的概念和相关算法以及面向对象的核心思想是学习Java的“主旋律”。有效地巩固、掌握Java这一主旋律也正是作者编写本书的目的。

本书主要内容是通过有趣的故事或场景，帮助读者加深对Java程序设计中涉及的一些核心概念、面向对象的思想和重要的设计模式的理解，进一步巩固教材的学习效果。本书按照场景故事、场景故事的目的、程序运行效果与视频讲解以及阅读源代码的模式，涵盖表达式与语句、类与对象、子类与继承、接口与实现、内部类与异常类、Lambda表达式、常用实用类、线程、集合框架、输入输出流、GUI程序设计、播放音频、绘制图形图像、面向抽象、接口编程的基本思想和部分设计模式（如策略模式、访问者模式、装饰模式、责任链模式）等知识点。

本书特色

（1）幽默风趣的场景故事：本书用言简意赅的场景故事，最大限度地体现Java编程的概念、思想或方法，以帮助读者加深对知识点的理解和记忆。本书的部分场景故事是众所周知的，部分场景故事完全是作者虚构的。

（2）场景故事的目的：恰如其分地启发思考，明确学习目标。由侧重点、涉及的其他知识和进一步尝试这三部分构成。

“侧重点”旨在让读者掌握此场景故事所侧重的知识、方法或思想。

“涉及的其他知识点”是相对于侧重点而言，由于本书属于参考教材，在内容的难度和广度上大于主教材，因此每个专题除了有侧重的知识和方法外，还会涉及一些其他知识点。

“进一步尝试”用于提示读者进一步需要思考的问题。

（3）知识体系由浅入深，重要知识点突出，可碎片化阅读。

本书内容按照由浅入深的知识体系展开，但不限于按部就班的、枯燥的Java知识体系，每个“场景故事”有自己的侧重知识点，这些知识点也包含在Java程序设计的相关教材中。

本书在力求突出重要知识点的同时，努力做到给读者带来趣味的、有深度和广度的阅读体验，以达到帮助读者加深和巩固主教材所学的知识点，扩展学生的知识面。

对于正在学习Java教材的高校学生，可以根据学习进度适当阅读本书中的趣味故事，以加强教材的学习效果。在学习完教材的主要内容之后，可再翻阅本书，也一定会有更深的体会和知识掌握程度的提高。

（4）配套资源丰富。

为便于学习，本书配有教学视频、源代码等资源。

① 430分钟视频讲解：突出重要概念、算法和设计理念的讲解，手把手带你学会实操。

获取教学视频方式：读者可以先扫描本书封底的文泉云盘防盗码，再扫描书中相应的视频二维码，观看教学视频。

② 源代码：阅读源代码可以让读者全面理解场景故事要体现的核心知识点、

算法或重要的编程思想。

获取源代码方式：先扫描本书封底的文泉云盘防盗码，再扫描下方二维码，即可获取。

源代码

本书可以作为作者主编的《Java 2实用教程（第5版）》（ISBN：9787302464259）和《Java面向对象程序设计（第3版）-微课视频版》（ISBN：9787302540526）的参考教材，也可以作为其他Java程序设计相关教材的参考教材。

作　者

2021年2月

CONTENTS 目录

一举两得

1.1 场景故事

Tom初学Java，从Java的第一个程序中学会了在命令行输出Hello World：

```
System.out.println("Hello World");
```

Tom听老师说过：

“初学Java时，务必掌握使用Java SE提供的JDK调试程序等知识。如果学生能够熟练地使用JDK，就可以再选一个流行的或自己喜爱的IDE，而且大部分IDE都是相似的，只要有了JDK基础，很快就能学会。”

有一天，经常和Tom一起上公选课的Jerry羡慕地对Tom说：“哇！你竟然会用Java编程写程序。”于是，Jerry想请Tom帮忙编写一个程序，来解决他最近遇到的问题。Jerry对Tom说：“最近我经常需要计算一些数的平均数，计算起来好麻烦，你能帮帮我吗？我把数据读给你，然后你把计算结果告诉我，可以吗？”

Tom想了想自己已经学会的Java知识，觉得自己可以帮助他，就答应了Jerry。

第二天，Tom写好了程序，两人就开始计算平均数。每计算出一个平均数，Tom就在命令行上右击，复制命令行显示的结果，然后把结果粘贴到一个用“记事本”程序打开的文件中，如图1.1所示。

过了一会儿，Tom突然对Jerry说：“不好了，刚才有一次，我好像忘记把结果粘贴到记事本上了！” Tom一下子不知所措，因为Tom也弄不清楚具体是哪一次没有把结果粘贴到记事本上，而且Tom还曾用cls命令清过屏。Tom对Jerry说：“我回去改进一下程序，星期三咱俩再继续算吧。”

到了星期三，Jerry和Tom又开始计算平均数。Jerry问Tom：“你这次不会忘记复制粘贴结果了吧？” Tom得意地说：“这次不用复制粘贴了，因为我加了几行代码，可以实现在命令行上显示结果的同时，在一个文本文件里自动地记录计算结果！”

Jerry对Tom说；“哇，你好厉害，这不就是一举两得嘛！”

图1.1　Tom和Jerry在计算平均数

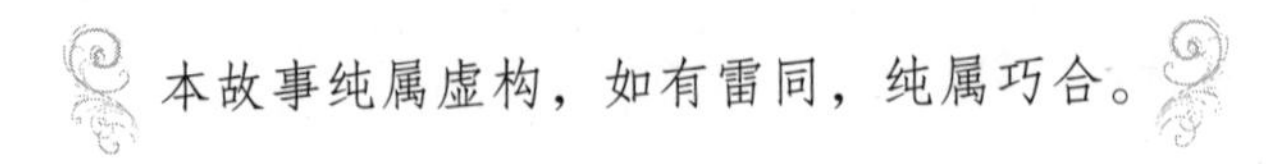

1.2　场景故事的目的

1. 侧重点

out是System类中的static成员变量，类型是PrintStream类型的输出流（一个PrintStream类的对象）。System类提供的这个默认的out输出流的目的地是命令行，所以out.print()、out.println()或out.printf()就会把数据显示在命令行。System类可以用类名调用，例如：

```
public static void setOut(PrintStream out);
```

方法可以重新设置out输出流的目的地。如果用户希望out输出流的目的地是一个文件，如record.txt文件，就可以进行如下设置：

```
PrintStream saveOut = Syetem.out;
PrintStream outToFile =new PrintStream(new File("record.txt"));
System.setOut(outToFile);
```

如果想让out输出流的目的地再切换到命令行，则执行如下代码：

```
System.setOut(saveOut);
```

通过不断地更换out输出流的目的地，就可实现“一举两得”，即把数据显示在命令行的同时也把数据写入record.txt文件。

2. 涉及的其他知识点

基本类型数据，循环语句，从键盘输入基本类型数据。

3. 进一步的尝试

将每次计算过程的结束时间或完成人显示在命令行，同时也保存到文本文件。

1.3 程序运行效果与视频讲解

在命令行显示数据的同时，也将数据保存到某个.txt文件中。主类是MainClass。程序运行效果如图1.2（a）和图1.2（b）所示。

视频讲解

```
(继续)计算请输入true,否则输入false
true
用空格做分隔输入若干个数,然后用空格、字符结束，回车确认
56 89 100 256 89 a
5个数的和为590.0
5个数的平均值118.0
(继续)计算请输入true,否则输入false
true
用空格做分隔输入若干个数,然后用空格、字符结束，回车确认
0.618 3.14 2.787  5.16 w
4个数的和为11.705
4个数的平均值2.92625
(继续)计算请输入true,否则输入false
```

(a) 在命令行显示数据

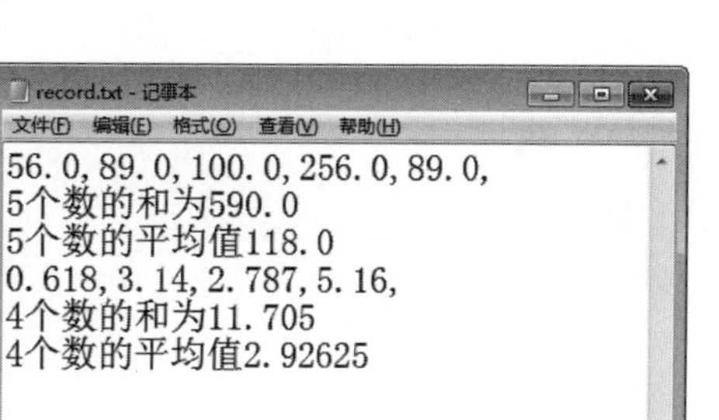

(b) 数据同时存入.txt文件

图1.2 命令行与记事本

1.4 阅读源代码

MainClass.java的代码如下：

```
import java.util.Scanner;
import java.io.*;
public class MainClass {
    public static void main (String args[ ]){
        PrintStream saveOut = System.out;
        PrintStream outToFile =null;
        try {
           outToFile =new PrintStream(new File("record.txt"));
```

```
        }
        catch(FileNotFoundException exp) {}
        boolean isContinue = true;
        double sum=0;
        int count=0;
        while(isContinue) {
           System.out.println("(继续)计算请输入true,否则输入false");
           Scanner getBoolean=new Scanner(System.in);
           isContinue = getBoolean.nextBoolean();
           Scanner reader=new Scanner(System.in);
           System.out.println("用空格做分隔输入若干个数,然后用空格、字符结
           束，回车确认");
           while(reader.hasNextDouble()){
              double x = reader.nextDouble();
              System.setOut(outToFile);
              System.out.print(x+",");
              count++;
              sum=sum+x;
           }
           System.setOut(saveOut);
           System.out.println(count+"个数的和为"+sum);
           System.out.println(count+"个数的平均值"+sum/count);
           System.setOut(outToFile);
           System.out.println("");
           System.out.println(count+"个数的和为"+sum);
           System.out.println(count+"个数的平均值"+sum/count);
           System.setOut(saveOut);
           count = 0;
           sum = 0;
        }
    }
}
```

精准的天平

2.1 场景故事

一家名字是double的玩具工厂，其产品上标识的产品重量以d或D作为后缀，如2.89D、0.4d，而且这家工厂很有名气，经常省略后缀d。但即使double玩具工厂的产品标识上省略后缀d，大家也都知道这是double工厂生产的玩具产品。

后来，又有一家名字是float的玩具工厂，规模没有double工厂大，其产品上标识的产品重量以f或F作为后缀，如2.89F、0.4f，但不允许float工厂省略后缀f或F。

一天，某顾客分别在double工厂和float工厂买了两个外观一样、显示同样重量的玩具。这名顾客好奇地将两个玩具放在天平的两端，他惊讶地发现天平出现了倾斜，如图2.1所示。float工厂的玩具竟然比double工厂的玩具重一些。顾客感觉很迷惑，这是为什么呢？

图2.1　精准的天平

顾客去询问double工厂的技术人员，技术人员答复说：“double工厂的玩具的标识虽然是0.4或0.4d，但产品的实际重量非常接近数学意义上的0.4，即误差很小。而float工厂的玩具的标识是0.4f，但由于float工厂的实现手段有限，其产品的实际重量和数学意义上的0.4的误差比double工厂的大。不是所有产品标识的重量和数学意义上的数字都有误差，例如0.125、0.75等标识就没有任何误差。如果您购买的我们两家的产品上的重量标识分别是0.125d和0.125f，那么把两件产品放在天平的两端，天平就是绝对平衡的。”接着，技术人员又向顾客介绍了IEEE二进制浮点数算术标准（IEEE 754），顾客终于明白了其中的道理。

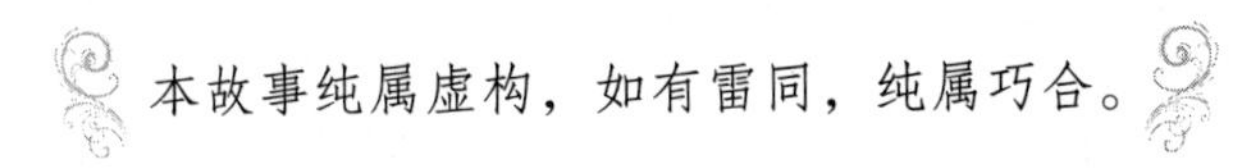

2.2 场景故事的目的

1. 侧重点

进一步理解double和float精度的不同之处。正整数的二进制是“商除以2求余”直到商为0，纯小数的二进制是通过“纯小数部分乘2取整”直到整数部分为0。与整数的二进制不同，纯小数的二进制可能有无限多位。例如：

$0.4 \times 2 = 0.8$
$0.8 \times 2 = 1.6$
$0.6 \times 2 = 1.2$
$0.2 \times 2 = 0.4$
$0.4 \times 2 = 0.8$
$0.8 \times 2 = 1.6$
……

那么，0.4的二进制就是：

0.0110 0110 0110 0110……

即0.4的二进制有无限多位。按照数学计算，有下列等式成立：

$0.4 = 0 \times 2^{-1} + 1 \times 2^{-2} + 1 \times 2^{-3} + 0 \times 2^{-4}$ ……

那么，截取上面等式右端无穷级数中的任何有限项的代数和，都是0.4的近似值，这就意味着截取的项越多，精度越高。

按照IEEE 754，首先移动小数点（以二进制表示），保证整数部分只有1位非零的1（浮点数的来历是移动小数点）。对于0.4，就是小数点向右移动两位：

1.10011001100110011001100110 0110 ...

那么，按照IEEE 754，指数是-2（从数学的角度来看，后者是前者乘以2的-2次幂得到的）。float精度是4字节，占32位。

按照IEEE 754，0.4f的存储是：

（1）符号位：0存在第31位上（正数存0，负数存1）。

（2）指数：-2+127=125，占8位（即第 30~23位），存放的是125的二进制01111101（指数多加一个偏移量127是技术上的需要）。

（3）尾数（第 22~0 位）就是从

1.10 0110 0110 0110 0110 0110 0110 0110 01

中小数点后面取23位（第24位上的数字需要“1进0舍”，进位到第23位），尾数是：

10 0110 0110 0110 0110 0110 1

0.4f的IEEE 754存储如图2.2所示。

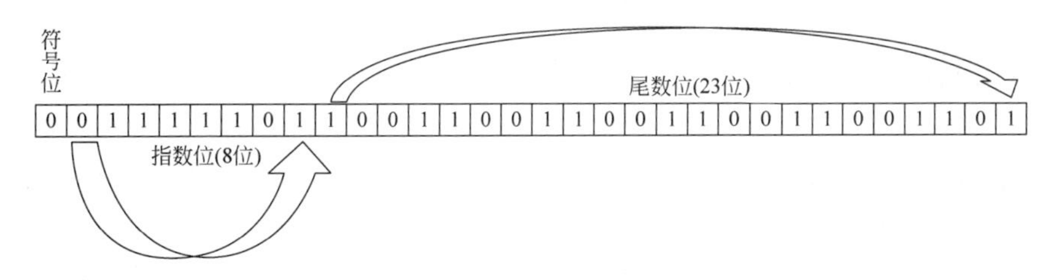

图2.2　0.4f的IEEE 754存储

注意，转化为十进制（按照IEEE 754标准，假如保留小数点后20位），是：

0.40000000596046450000

2. 涉及的其他知识点

Integer类以及Long类调用static方法得到float型浮点数和double型浮点数的IEEE 754的标准表示。

3. 进一步的尝试

比较0.125d和0.125f的大小，看看二者的IEEE 754的标准表示。

2.3 程序运行效果与视频讲解

视频讲解

Java程序在输出正浮点数的IEEE 754 的标准表示时，如果符号位是0（即正浮点数），则不输出符号位上的0；如果符号位是0（即正浮点数）并且指数也是正数，则省略输出指数上的符号位。比如，对于正的float型浮点数，输出IEEE 754的标准表示往往是30或31位；对于正的double型浮点数，输出的往往是62或63位。这里的程序输出了-0.4d和-04f的IEEE 754的标准表示。主类是DoubleAndFloat。程序运行效果如图2.3所示。

```
false
true
10111110110011001100110011001101
1011111111011001100110011001100110011001100110011001100110011010
-0.40000000596046450000
-0.40000000000000000000
```

图2.3 double型浮点数与float型精度比较

2.4 阅读源代码

DoubleAndFloat.java的代码如下：

```
public class DoubleAndFloat {
   public static void main(String args[]) {
      float x = -0.4f;     //x的值模拟float玩具厂的玩具实际重量
      double y = -0.4;     //y的值模拟double玩具厂的玩具实际重量
      System.out.println(x==y);
      System.out.println(x<y);
      System.out.println(Integer.toBinaryString(Float.floatToIntBits(x)));
      System.out.println(Long.toBinaryString(Double.doubleToLongBits(y)));
      System.out.printf("%20.20f\n",
      Float.intBitsToFloat(Float.floatToIntBits(x)));
      System.out.printf("%20.20f\n",
      Double.longBitsToDouble(Double.doubleToLongBits(y)));
   }
}
```

点名与大奖

3.1 场景故事

小明上课的教室里一共有两个班，其中一班有30名同学、二班有33名同学。小明发现老师每次按学号分别在两个班点名6位不同的同学，并且老师每次都是简单地看一下她的计算机，就能快速、随意地说出6个不同的学号。老师点名现场如图3.1所示。

图3.1 老师点名现场

课程快结束时，小明好奇地问老师："老师，我发现您的记忆力非常好。您可以在一班或二班快速、随意地说出6个学号，而且能保证6个学号互不相同。若是您想要点名20名学生，即快速、随意地说出20个学号，也能保证这20个学号是不同的吗？您是怎么做到的呢？"

老师笑着对小明说："我每次点名时，都是用余光看着计算机显示器的屏幕，是我用自己编写的一个点名程序，来保证从班级里抽取6个不同的学号 。"

小明向老师要了源代码，决定认真研究一下，准备写一个模拟福利彩票抽奖的程序。小明发现，这学期有12位给自己上课的老师，决定分别给这些老师编号为1~12。如果是10号老师上课，就设篮球号码是10；然后用自己编写的Java程序随机得到1~33中的6个数，作为红球号码，再与福利彩票开奖后的大奖做比对，看看自己的Java程序能否得大奖。

本故事纯属虚构，如有雷同，纯属巧合。

3.2 场景故事的目的

1. 侧重点

java.util包（Java工具包）中的Arrays类提供了许多static方法，可以用类名Arrays直接调用这些方法。

通过该场景学习Arrays类提供的复制数组和判断数组相等的方法。

（1）数组的复制。

```
public static int[] copyOfRange(int[] original, int from, int to)
```

copyOfRange()方法可以把参数original指定的数组中从索引from至to-1的元素复制到一个新数组中，并返回这个新数组，即新数组的长度为to-from。如果to的值大于数组original的长度，则新数组的元素从索引original.length-from开始的元素取默认值。

比如，对于数组a：

```
int [] a = {1,2,3,4,5,6};
```

① 复制数组a的部分元素，得到一个新的数组b：

```
int [] b = Arrays.copyOfRange(a,1,4);
```

那么，数组b的长度即b.length的值是3。b[0]的值是2, b[1]的值是3, b[2]的值是4。

② 复制数组a的部分元素，得到另一个新的数组：

```
int [] c = Arrays.copyOfRange(a,4,10);
```

那么，数组c的长度即c.length的值是6。c[0]的值是5，c[1]的值是6，c[2]至c[5]的值都是0。

（2）数组的相等判断。

有时需要判断两个类型相同的数组的值是否相同，即二者的长度一样且对应的元素值也相同。

public static boolean equals(int[] a,int[] b)方法可以判断int型数组a和数组b是否相等（类似的方法还有double、char、byte型数组的equals()方法）。当数组a和数组b相等时，该方法返回true；否则返回false。

比如，在编写标准化考试的程序时，将标准答案（多选题）如ABD存放在一个数组里，代码如下：

```
char [] answer = {'A','B','D'};（也许某个开发者写成{'B','A','D'};）
```

考试者即用户提交的答案数组是：

```
char [] user = {'B','D','A'};
```

或

```
char [] user = {'B','A'};
```

那么，怎样使用equals()方法给用户判定成绩呢？换言之，在某些应用中，只要两个数组长度一样，含有的数据也一样，就认为二者相等，而不要求两个数组的元素按索引顺序依次相同。一个简单的办法就是把二者都排序，代码如下：

```
Arrays.sort(answer);
Arrays.sort(user);
```

然后再使用boolean equals(char []a,char[] b)方法，代码如下：

```
boolean isRight = Arrays.equals(answer,user);
```

得到一个值isRight，再根据isRight是true或false判断用户提交的数组和标准答案的数组是否相等，然后决定给多少分。

2．涉及的其他知识点

java.util包中的Random类以及for循环语句。

3．进一步的尝试

从不连续的若干个数字中随机抽取若干个数。例如，从1、2、5、8、10、14、23、30中随机抽取6个数字。

3.3 程序运行效果与视频讲解

视频讲解

老师分别在一班和二班随机提问6名同学。主类是ExtractNumber。程序运行效果如图3.2所示。

```
一班被抽到的回答问题的学号:
[28, 12, 14, 11, 26, 20]
二班被抽到的回答问题的学号:
[33, 11, 9, 20, 16, 10]
两个班抽取的学号相同吗？false
```

图3.2 点名与大奖的程序运行效果

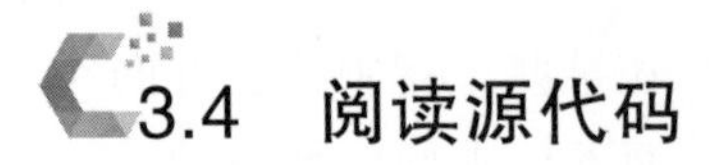

3.4 阅读源代码

（1）GetRandomNumber.java的代码如下：

```
import java.util.Random;                        //负责得到随机数
import java.util.Arrays;
public class GetRandomNumber {
   public static int [] getRandomNumber(int max,int amount) {
       int [] result = new int[amount<max?amount:max];
                                                //存放得到的amount随机数
       int [] a = new int[max];                 //存放max个整数
       Random random = new Random();
       for(int i = 0;i<a.length;i++){
                                                //将1至max个数放入数组a
           a[i] = i+1;
       }
       for(int i =0;i<amount;i++) {             //得到amount个随机数
          int index  = random.nextInt(a.length);
                                                //随机得到数组的一个索引值
          result[i] = a[index];
          int [] b = Arrays.copyOfRange(a,0,index);
                                                //数组b存放的是a[0]至a[index-1]
          int [] c = Arrays.copyOfRange(a,index+1,a.length);  //数组复制
          a = new int[b.length+c.length];
                                                //新的数组a中去掉了抽到的数字
```

```
            for(int k=0;k<a.length;k++) {
               if(k<b.length)
                  a[k] = b[k];
               else
                  a[k] = c[k-b.length];
            }
         }
         return result;
   }
}
```

（2）ExtractNumber.java的代码如下：

```
import java.util.Arrays;
public class ExtractNumber {
   public static void main(String args[]) {
         int [] a =GetRandomNumber.getRandomNumber(30,6);
         System.out.println("一班被抽到的回答问题的学号：");
         System.out.println(Arrays.toString(a));
         int [] b =GetRandomNumber.getRandomNumber(33,6);
         System.out.println("二班被抽到的回答问题的学号：");
         System.out.println(Arrays.toString(b));
         Arrays.sort(a);
         Arrays.sort(b);
         boolean isSame =Arrays.equals(a,b);
         System.out.println("两个班抽取的学号相同吗？"+isSame);
   }
}
```

场景故事 4

爱情故事

4.1 场景故事

巫婆用巫术把男孩心仪的女孩藏到了纵横交错、高度各不相同的群山中的最高山上。女孩所在的位置是高度为50的山上，如图4.1所示。

0	21	14	40	46	28	48	38	30	11
45	42	32	12	33	34	5	1	50	9
37	8	20	19	22	43	4	2	49	3
24	6	41	39	18	10	31	29	16	23
27	17	36	44	35	13	47	25	15	7

图4.1　纵横交错、高低不同的山

男孩多日不见女孩，万般焦急。于是他决定开始寻觅女孩，打算从图4.1中左上角的山出发。在这个时候，幸运小精灵路过此地，见此情形，心生善意，便告诉男孩一个救回女孩的办法。

幸运小精灵对男孩说："你可以以其中的一座山作为出发点，首先环视当前山的东、西、南、北4个方向的山，因为巫婆的巫术使你无法看见更远的山。然后，必须在其中找到一座最高的山爬上去，否则巫婆会藏起所有的山，你就再也找不到女孩了。另外，建议你每爬一座山就留下一个标记，即记住自己曾爬过的山。也许在某个时候，你会发现当前山的东、西、南、北4个方向的山都是曾爬过的山了。这个时候，你只要按照曾留下的标记，一步一步地退回、一直退到曾爬过的某个山，发现此座山的东、西、南、北4个方向的山中，除了曾爬过的山以外，还有没爬过的山，那么就在这些山中再选最高山去爬，如此循环往复。只要无畏艰险，你一定能救回心仪的女孩。" 男孩相信了幸运小精灵的话，最终救回了心仪的女孩。

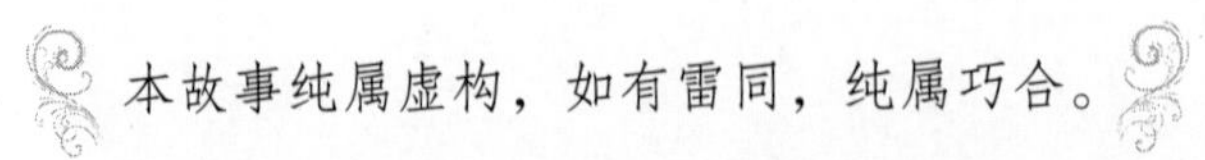

4.2 场景故事的目的

1. 侧重点

Java语言的控制语句、循环语句和C语言非常类似，学习Java时只注意Java和C的不同之处即可，如条件表达式的值必须是boolean类型。

2. 涉及的其他知识点

数组以及java.util包中的Stack、ArrayList和Collections类（这些均属于集合框架的知识点）。

数组属于引用型变量。两个相同类型的数组如果具有相同的引用，它们就有完全相同的元素。编译时，不检查数组索引是否越界。但当程序运行时，一旦发现数组索引越界，就会触发ArrayIndexOutOfBoundsException异常。

Stack<E>实现了泛型接口List<E>，Stack<E>泛型类的对象采用栈式结构存储数据（先进后出），习惯上将Stack类创建的对象称为堆栈。使用Stack<E>泛型类声明创建堆栈时，必须要指定E的具体类型，然后堆栈就可以使用void push(E item)实现压栈操作，使用public E pop()方法实现弹栈操作。

ArrayList<E>实现了泛型接口List<E>，ArrayList<E>泛型类的对象采用顺序结构来存储数据。通常将ArrayList类创建的对象称为数组表。当使用ArrayList<E>泛型类声明创数组表时，必须要指定E的具体类型，然后数组表就可以使用add(E obj)方法向数组表依次增加节点。

Collections类还提供了将List接口的数据重新随机排列的类方法：public static void shuffle(List<E> list) 将list中的数据按洗牌算法重新随机排列。

3. 进一步的尝试

输出男孩曾爬过的山的高度总和，包括回退时爬过的高山

【提示】用全部山的高度之和，减去男孩不曾爬过的山的高度之和。

输出男孩曾爬过的高山的高度总和，但不包括回退时爬过的高山。

4.3 程序运行效果与视频讲解

视频讲解

由于每次运行程序时，纵横交错、高度各不相同的二维数组的元素的值都有变化，因此每次运行效果不尽相同。

效果图中的圆圈是男孩寻觅的起点，实心三角表示男孩寻觅过程中曾爬过的山，空心三角表示不曾爬过的山，实心五角星是女孩所在的山。主类是SaveTheGirl。程序运行效果如图4.2所示。

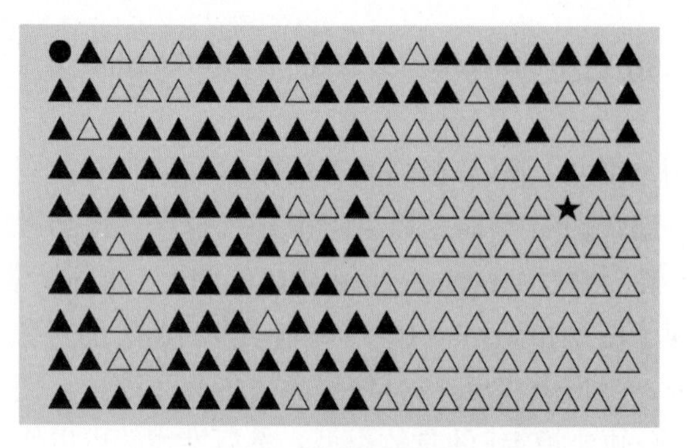

图4.2 爱情故事的程序运行效果

4.4 阅读源代码

SaveTheGirl.java的代码如下：

```
import java.util.ArrayList;
import java.util.Stack;
import java.util.Collections;
public class SaveTheGirl {        //代码中打◆号注释的语句，暂时不要求看懂
    public static void main(String args[]) {
        Stack <Integer> saveI = new Stack <Integer>();
                                  //◆存放爬过的山的索引坐标i
        Stack <Integer> saveJ = new Stack <Integer>();
                                  //◆存放爬过的山的索引坐标j
        ArrayList<Integer> list = new ArrayList<Integer>();
                                  //◆存放整数的数组表
        int row = 10;             //数组的行数
        int column = 20;          //数组的列数
```

```
for(int i=1;i<=row*column;i++) {
                              //把1~row*column个数放入链表
   list.add(i);
}
Collections.shuffle(list);           //◆打乱数组表，即洗牌
int [][] a = new int[row][column]; //二维数组，刻画纵横交错的山
int index = 0;
for(int i = 0;i<row;i++) {
   for(int j = 0;j<column;j++){
      a[i][j] = list.get(index); //得到数组表的第index个节点的
                                 //值，设置山的高度
      index++;
   }
}
a[0][0] = 0;                         //男孩寻觅的起点
int MAX = row*column;                //最高山的高度（要找的数据）
int posionX = 0;
int posionY =0;                      //将来存放找到的女孩（数据）的位置
//男孩寻觅的过程，只要a[i][j]的值被设置成0，就表示a[i][j]是男孩爬过的山
int max = -1;
int i = 0;
int j = 0;
int m =0,n =0;
while(true){
   saveI.push(i);                    //记下爬过的山的位置(压栈)
   saveJ.push(j);                    //记下爬过的山的位置
   boolean boo =                //判断东西南北周围的山是否都爬过了啊
   a[i][j+1<column?j+1:column-1]==0&&
                                     //判断是否爬过a[i][j]的东侧山
   a[i][j-1>0?j-1:0]==0&&            //判断是否爬过西侧山
   a[i+1<row?i+1:row-1][j]==0&&  //判断是否爬过南侧山
   a[i-1>0?i-1:0][j]==0;             //判断是否爬过北侧山
   while(boo){  //假如a[i][j]的东、西、南、北4个方向的山都爬过了
       i= saveI.pop();
       j = saveJ.pop();              //退回一步
       boo =a[i+1<row?i+1:row-1][j]==0&&
       a[i-1>0?i-1:0][j]==0&&
       a[i][j+1<column?j+1:column-1]==0&&
       a[i][j-1>0?j-1:0]==0;
       //一直退到a[i][j]的东、西、南、北4个方向的山中有没爬过的山
   }
   //这样就可以找一个东、西、南、北4个方向中的高山爬
   if(i+1<row&&a[i+1][j]>=max&&a[i+1][j]!=0){
   //检查a[i][j]的北山的高度
```

```
            m=i+1;
            n =j;                    //保存该方向的索引到m、n中
            max = a[i+1][j];
        }
        if(i-1>=0&&a[i-1][j]>max&&a[i-1][j]!=0){
        //继续检查a[i][j]的南山的高度
            m=i-1;
            n =j;                    //保存该方向的索引到m、n中
            max = a[i-1][j];
        }
        if(j+1<column&&a[i][j+1]>max&&a[i][j+1]!=0){
        //继续检查a[i][j]东山的高度
            m=i;
            n =j+1;                  //保存该方向的索引到m、n中
            max = a[i][j+1];
        }
        if(j-1>=0&&a[i][j-1]>max&&a[i][j-1]!=0){
        //继续检查a[i][j]的西山的高度
            m=i;
            n =j-1;                  //保存该方向的索引到m、n中
            max = a[i][j-1];
        }
        if(max == MAX){
            posionX = m;
            posionY = n;             //终于找到了女孩，记下位置
            break;                   //立刻停止循环
        }
        a[m][n] = 0;                 //没找到，将a[m][n]设置成爬过的山
        i = m;
        j = n;
        max = -1;                    //走到a[m][n]，继续找下一个节点
    }
    //以下输出男孩爬过的山和没爬过的山
    for(int p=0;p<row;p++) {
        for(int q =0;q<column;q++) {
            if(a[p][q]==0) {
                if(p==0&&q==0)
                    System.out.print("●");   //男孩的起点
                else
                    System.out.print("▲");   //表示爬过的山
            }
            else {
                if(a[p][q]==MAX)
                    System.out.print("★");   //女孩的位置
```

```
                else
                    System.out.print("△");              //没爬过的山
            }
        }
        System.out.println();                           //换行
    }
    System.out.printf("\n女孩的位置:%d,%d",posionX,posionY);
                                                        //女孩的位置索引
  }
}
```

场景故事 5

石头与钻石

5.1 场景故事

很久以前，一位好心的农夫路遇一位落魄的老人，农夫就把身上的8个铜钱送给了这位老人。后来的某一天，农夫又遇见了他曾经帮助过的这位老人。

老人突然对他说："我这里有8块一模一样的小石头，你整理出一块长方形的农田，把它再划分成大小相等的一些小方块，然后把这8块石头分别放到你划分出的小方块里，但必须要保证在有小石头的方块中，每个方块只能放一块小石头。"

农夫问："在农田里放石头，又不会长出庄稼，你为何要让我这样做呢？"

老人说："这些小石头是很神奇的，你一旦把它们放入农田的小方块中，农田中的小方块就会慢慢地发生神奇的变化。"

农夫又问："会发生什么变化啊？"

老人答道："农田里没有放小石头的小方块，可能在第二天出现小石头或继续没有小石头；而放有小石头的小方块，到了第二天小石头可能消失了或仍有小石头。你的农田中每个小方块的邻居就是它旁边的8个小方块，注意至多8个。"

然后，老人说出了变化的秘密：

如果一个小方块的邻居中有3个小方块里有小石头，第二天，该小方块里有小石头。

如果一个小方块的邻居中有两个小方块里有小石头，第二天，该小方块的状态保持不变，即如果原来小方块有小石头就仍有小石头，小方块没有小石头就仍没有小石头。

如果是其他情况，第二天，小方块里没有石头。

老人还在地上给农夫画了一个石头奥妙图，如图5.1所示。这个图很好地解释了其中的奥妙（变化规律）。

接着老人告诉农夫一件让农夫更为惊讶的事情，老人对农夫说："如果在50天的时间里，农田里小方块的状态从某天开始不再发生变化，那么这些小石头就都会变成钻石。而且，小石头在农田里分布得越好看，变成的钻石的质量就越大！"

说完，老人便消失了。

图5.1　石头奥妙图

农夫每天都在琢磨怎样把小石头放入农田的小方块里。农夫很聪明，他知道农田里小方块的后续状态依赖于第1天的状态。农夫经过分析，认为几天之后农田会出现以下3种可能的情况。

① 农田里的小方块可能会进入一个稳定状态，即有部分石头不再随时间变化而变化。

② 农田可能在某一天进入一个周期状态，即在几个状态之间周而复始地循环往复。

③ 农田在某一天也许会进入一个非常糟糕的稳定状态，即农田里每个小方块里都没有石头了！

后来，农夫想出了一个放石头的方案，最后收获了40颗钻石。假如你是这位农夫，你能得到比他更多的钻石吗？

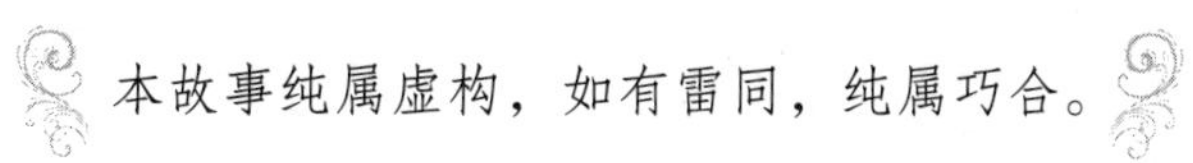

5.2 场景故事的目的

1. 侧重点

二维数组的复制和迭代的算法以及怎样使用控制和循环语句操作二维数组。

2. 涉及的其他知识点

复制数组的方法代码如下：

```
public static int[] copyOf (int[] original)
```

该方法可以把参数original指定的一维int型数组的单元的值复制到一个新的一维int型数组中，并返回这个数组的引用，即新数组的长度等于数组original的长度，各个单元的值也和数组original的完全相同。

3. 进一步的尝试

试输出农田里中小石头变化过程中的某个中间状态，如输出第5天和第12天的状态。

5.3 程序运行效果与视频讲解

视频讲解

从一个位置开始纵向放置8个小石头。主类是StoneAndDiamond。程序运行效果如图5.2（a）和图5.2（b）所示。

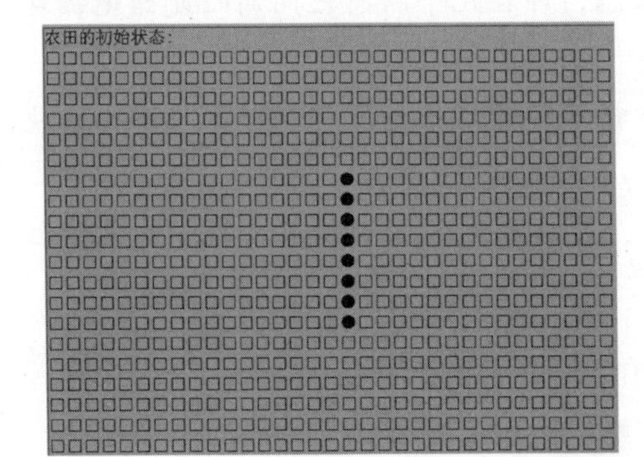

(a) 农田的初始状态

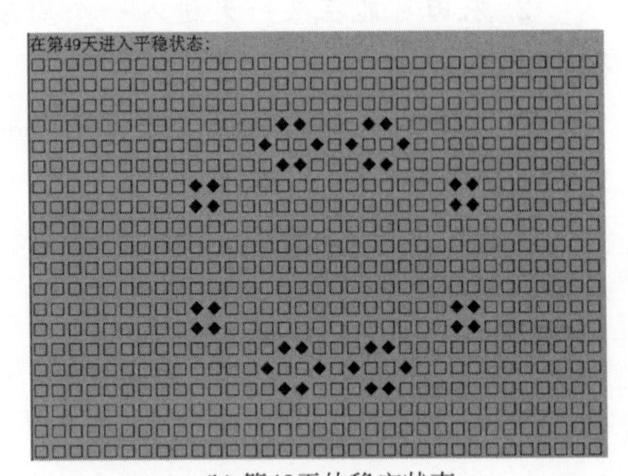

(b) 第49天的稳定状态

图5.2 石头与钻石的程序运行效果

5.4 阅读源代码

StoneAndDiamond.java的代码如下：

```
import java.util.Arrays;
public class StoneAndDiamond {
    public static void main(String args[]){
        int [][] farmland = new int[20][33];          //刻画农田状态的二维数组
        int maxDays = 50;                             //看50天内农田的变化
        for(int i=6;i<=13;i++){
             farmland[i][17] = 1;                     //1表示方块里有石头
        }
        System.out.println("农田的初始状态:");        //输出初始状态
        for(int i=0;i<farmland.length;i++ ) {
                for(int j = 0;j<farmland[i].length;j++){
                    if(farmland[i][j] == 1)
                      System.out.printf("%s","●"); //表示有石头
                    else
                      System.out.printf("%s","□"); //表示没有石头
                }
                System.out.println();
        }
        int [][] copyOfFarmland = new int[farmland.length][];
        for(int m = 0;m<copyOfFarmland.length;m++) {
             //将farmland复制到数组copyOfFarmland
             copyOfFarmland[m] = Arrays.copyOf(farmland[m],farmland[m].
            length);
        }
        int day = 1;
        while(day<=maxDays) {
           for(int i=0;i<copyOfFarmland.length;i++ ) {
               for(int j = 0;j<copyOfFarmland[i].length;j++){
                    //检查copyOfFarmland[i][j]周围活的石头个数
                    int stoneCounts  = 0;              //石头个数
                    if(i<copyOfFarmland.length-1) {
                    if(copyOfFarmland[i+1][j] == 1)    //检查当前小方块的正下方
                          stoneCounts++;
                    }
                    if(i>=1) {
                    if(copyOfFarmland[i-1][j] == 1)    //检查当前小方块的正上方
                          stoneCounts++;
                    }
                    if(j<copyOfFarmland[i].length-1) {
                     if(copyOfFarmland[i][j+1] == 1)   //检查当前小方块的正右方
                          stoneCounts++;
                    }
                    if(j>=1) {
                     if(copyOfFarmland[i][j-1] == 1)   //检查当前小方块的正左方
```

```
                stoneCounts++;
            }
            if(i<copyOfFarmland.length-1&&j<copyOfFarmland[i].
            length-1){
                if(farmland[i+1][j+1] == 1) //检查当前小方块的右下方
                stoneCounts++;
            }
            if(i<copyOfFarmland.length-1&&j>=1){
                if(farmland[i+1][j-1] == 1)   //检查当前小方块的左下方
                stoneCounts++;
            }
            if(i>=1&&j>=1){
                if(copyOfFarmland[i-1][j-1] == 1)
                                        //检查当前小方块的左上方
                stoneCounts++;
            }
            if(i>=1&&j<copyOfFarmland[i].length-1){
                if(copyOfFarmland[i-1][j+1] == 1)
                                        //检查当前小方块的右上方
                stoneCounts++;
            }
            if( stoneCounts == 3){
                farmland[i][j] = 1;         //方块里有石头
            }
            else if(stoneCounts == 2){
                //保持不变
            }
            else {
                farmland[i][j] = 0;
            }
        }
    }
    //判断农田是否进入稳定状态
    boolean isStable = true;
    for(int m = 0;m<copyOfFarmland.length;m++) {
        if(!Arrays.equals(farmland[m],copyOfFarmland[m])) {
            isStable = false;
            break;
        }
    }
    if(isStable) {
        break;                          //结束while循环语句
    }
    day++;                              //继续执行while循环语句到下一天
```

```
            for(int m = 0;m<copyOfFarmland.length;m++) {
                //将farmland复制到数组copyOfFarmland
                copyOfFarmland[m] = Arrays.copyOf(farmland[m],farmland[m].
                length);
            }
        }
        if(day<=maxDays){          //存在稳定状态，输出稳定状态的样子
            System.out.println("在第"+day+"天进入稳定状态:");
            for(int i=0;i<farmland.length;i++ ) {
              for(int j = 0;j<farmland[i].length;j++){
                  if(farmland[i][j] == 1)
                    System.out.printf("%s","◆");       //表示农田里有钻石
                  else
                    System.out.printf("%s","□");
              }
              System.out.println();
            }
        }
        else {
           System.out.println("在"+maxDays+"天内没进入稳定状态,目前状态是: ");
            for(int i=0;i<farmland.length;i++ ) {
              for(int j = 0;j<farmland[i].length;j++){
                  if(farmland[i][j] == 1)
                    System.out.printf("%s","●");
                  else
                    System.out.printf("%s","□");
              }
              System.out.println();
            }
        }
    }
}
```

场景故事 6

守株待兔

6.1 场景故事

相传在古代，有一个农民，日出而作，日落而息。遇到好年景，也不过刚刚吃饱穿暖；一遇灾荒，就要忍饥挨饿了。他想改善生活，但他太懒，胆子又特小，总想得到主动送上门来的意外之财。

深秋的一天，奇迹终于发生了。他正在田里耕地，恰逢周围有人在打猎。吆喝之声四处起伏，受惊的小野兽拼命地奔跑。突然，有一只兔子，不偏不倚地一头撞在他旁边的树上……这就是后来人们相传的守株待兔的故事，如图6.1所示。

图6.1　守株待兔

6.2 场景故事的目的

1. 侧重点

通过大家熟知的成语故事——守株待兔，来巩固对象的基本知识点。

（1）类是一种重要的数据类型，且允许开发者编写的一种数据类型。

（2）类声明的变量称为对象变量，简称为对象。

（3）对象变量本身负责存放引用值，以便可以访问分配给该对象的变量（实体），调用类提供的方法。

（4）两个类型相同的对象，一旦二者的引用值相同，二者就具有完全相同的变量（实体）。

2. 涉及的其他知识点

了解java.util包中的Random，怎样使用Random随机获取一个boolean数据。

3. 进一步的尝试

在Rabbit类中增加修改weight属性和获取weight属性值的方法，使得Rabbit对象可以修改自己的weight属性值。

6.3 程序运行效果与视频讲解

在兔子开始奔跑之后，守株待兔的人开始所谓的守株待兔。主类是MainClass。程序运行效果如图6.2所示。

视频讲解

```
兔子没撞树...继续守株待兔...
兔子撞树上了，哈哈！
兔子撞死了，可以吃兔子肉了
兔子9.00kg
兔子撞树上了，哈哈！
兔子还活着，可以养兔子了！
兔子原来9.00kg
兔子现在9.80kg
```

图6.2 守株待兔的程序运行效果

6.4 阅读源代码

（1）Rabbit.java的代码如下：

```
import java.util.Random;
```

```
public class Rabbit {
    double weight = 9;    //兔子的质量，设置初始值9
    boolean isHitTree;    //是否撞树上，默认值false
    boolean isDie;
    public void rabbitHitTree(){
        isHitTree = true;
        Random random = new Random();
        isDie = random.nextBoolean();
    }
}
```

（2）People.java的代码如下：

```
public class People {
    Rabbit rabbitOfPeople = null; //组合Rabbit对象
    public void raiseTheRabbit() {
       if(rabbitOfPeople!=null){
          if(!rabbitOfPeople.isDie) {
             System.out.printf("\n兔子还活着,可以养兔子了额！");
             System.out.printf("\n兔子原来%.2fkg",rabbitOfPeople.weight);
             rabbitOfPeople.weight = rabbitOfPeople.weight+0.8;
              //兔子质量增加0.8
              System.out.printf("\n兔子现在%.2fkg",rabbitOfPeople.weight);
          }
          else {
             System.out.printf("\n兔子撞死了,可以吃兔子肉了");
             System.out.printf("\n兔子%.2fkg",rabbitOfPeople.weight);
          }
       }
       else {
           System.out.printf("\n没有兔子\n");
       }
    }
    public void waitForTheRabbit(Rabbit rabbit){ //守株待兔
       if(rabbit.isHitTree) {
           rabbitOfPeople = rabbit;
           System.out.printf("\n兔子撞树上了,哈哈！");
       }
       else {
           System.out.printf("\n兔子没撞树...继续守株待兔...);
       }
    }
}
```

（3）MainClass.java的代码如下：

```
public class MainClass {
    public static void main(String args[]) {
        People zhangSan= new People();
        Rabbit rabbit = new Rabbit();
        zhangSan.waitForTheRabbit(rabbit);//守株待兔
        rabbit.rabbitHitTree();
        zhangSan.waitForTheRabbit(rabbit);
        zhangSan.raiseTheRabbit();
        rabbit = new Rabbit();
        rabbit.rabbitHitTree();
        zhangSan.waitForTheRabbit(rabbit);//守株待兔
        zhangSan.raiseTheRabbit();
    }
}
```

新调虎离山

7.1 场景故事

从前有一座山，山上既没有猴子也没有老虎。

某天，一只猴子来到山中，并立刻声称：“山中无老虎，猴子称霸王。”

过了几天，一只老虎进山，吓得猴子立刻不敢作声了。

老虎声称：“我是山大王。”

某一天，山下突发事件，吸引老虎离山而去。猴子认为老虎凶多吉少，猴子马上声称：“山中无老虎，猴子称霸王。”

不料，没过几日，人们放虎归山，猴子立刻又不敢作声。

老虎继续声称：“我是山大王。”调虎离山的故事，如图7.1所示。

调虎离山

图7.1 调虎离山

本故事纯属虚构，如有雷同，纯属巧合。

7.2 场景故事的目的

1. 侧重点

巩固两个重要的基本知识点。

（1）两个类型相同的对象，一旦二者的引用值相同，二者就具有完全相同的变量（实体）。

（2）组合和参数传值的知识点。

2. 涉及的其他知识点

一个对象怎样利用null技术取消所组合的对象。

3. 进一步的尝试

在Mountain类中再多组合一些动物，比如Rabbit类的对象。当老虎离山时，让Rabbit类的对象说：“老虎离山了！”当老虎归山时，让Rabbit类的对象说：“老虎又回来了。”

7.3 程序运行效果与视频讲解

每当老虎不在山中时，猴子就说：“山中无老虎，猴子称霸王。”每当老虎进山后，老虎就说：“我是山大王。”主类是MainClass。程序运行效果如图7.2所示。

视频讲解

```
山中无老虎:
山中无老虎, 猴子称霸王,猴子38.90kg,高70cm
老虎进山:
我是山大王,我重108.67kg,高150cm
东北虎的质量108.670000
调虎离山:
山中无老虎, 猴子称霸王,猴子38.90kg,高70cm
放虎归山:
我是山大王,我重118.67kg,高150cm
放虎归山:118.67
```

图7.2　新调虎离山的程序运行效果

7.4 阅读源代码

（1）Monkey.java的代码如下：

```
public class Monkey {
    int height;
    float weight;
    public void monkeySpeak(){
        System.out.printf
        ("\n山中无老虎，猴子称霸王,猴子%.2f kg,高%d cm\n",weight,height);
    }
}
```

（2）Tiger.java的代码如下：

```
public class Tiger {
    double weight;                          //老虎的质量
    int height;                             //老虎的高度
    public void tigerSpeak(){
        System.out.printf("\n我是山大王,我重%.2f kg,高%d cm\n",weight,
        height);
    }
}
```

（3）MainClass.java的代码如下：

```
public class Mountain {
    Tiger tigerOfMontain;                   //组合Tiger对象
    Monkey monkeyOfMontain;                 //组合Monkey对象
    public void setTiger(Tiger tiger) {
       tigerOfMontain = tiger;
       tigerOfMontain.weight = tigerOfMontain.weight+10;
    }
    public void setMonkey(Monkey monkey) {
       monkeyOfMontain = monkey;
    }
    public void showMountainState(){
       if(tigerOfMontain == null) {
           monkeyOfMontain.monkeySpeak();
       }
       else {
          tigerOfMontain.tigerSpeak();
       }
    }
```

```
    public Tiger letTigerGoWay(){
        Tiger tiger = tigerOfMontain;
        tigerOfMontain = null;
        return tiger;
    }
}
```

场景故事 8

新击鼓传花

8.1 场景故事

新击鼓传花的游戏玩法：由若干个人（不少于两人）围成圆圈，首先让圆圈中的某人拿一束花，圈外有个背对着大家的人负责敲鼓（也可以敲击能发出声音的其他物体）；鼓响时，众人开始依次传花，直至鼓声停止为止。此时，花在谁的手中，谁就从圆圈中离开；然后，当鼓声再次响起时，圈中剩余的人继续依次传花。新击鼓传花游戏，如图8.1所示。

图8.1 新击鼓传花游戏

本故事纯属虚构，如有雷同，纯属巧合。

8.2 场景故事的目的

1. 侧重点

巩固组合的重要思想：一个对象如果想和另外一个对象发生联系，其方式就是组合后者。如果对象a组合了对象b，对象a就可以委托对象b调用方法产生行为，即对象a通过组合的办法复用对象b的方法（算法），俗称“为我所用”。

2. 涉及的其他知识点

了解怎样使用java.util.Random类获得一个随机数。例如，Random对象使用nextInt(8)返回0、1、2、3、4、5、6、7中的某个数，但不包括8。

如果People是类，那么People [] peopleInCircle = new People[10]；定义一个长度为10的People数组peopleInCircle。

这个数组peopleInCircle的每个单元用来存放People对象的引用，如：

```
peopleInCircle [0] = new People();
```

3. 进一步的尝试

在新击鼓传花过程中，当一个人从圆圈中退出时，程序输出此人把花传给了代号是多少的人。

8.3 程序运行效果与视频讲解

对象Flower负责从0、1、2、3、4、5、6、7中随机得到一个数。当花在某人手里，而此时对象Flower得到的随机数刚好是6（模拟鼓声停止），此人从圆圈中退出。主类是MainClass。由于程序每次的运行效果不尽相同，最后一人也不尽相同。程序运行效果如图8.2（a）和图8.2（b）所示。

视频讲解

```
代号7的人退出击鼓传花
代号3的人退出击鼓传花
代号6的人退出击鼓传花
代号2的人退出击鼓传花
代号8的人退出击鼓传花
代号1的人退出击鼓传花
代号4的人退出击鼓传花
代号12的人退出击鼓传花
代号11的人退出击鼓传花
代号9的人退出击鼓传花
代号5的人退出击鼓传花
代号10是最后一个人
```

(a) 10号是圈中最后一人

```
代号6的人退出击鼓传花
代号8的人退出击鼓传花
代号2的人退出击鼓传花
代号7的人退出击鼓传花
代号9的人退出击鼓传花
代号3的人退出击鼓传花
代号11的人退出击鼓传花
代号4的人退出击鼓传花
代号12的人退出击鼓传花
代号1的人退出击鼓传花
代号10的人退出击鼓传花
代号5是最后一个人
```

(b) 5号是圈中最后一人

图8.2 新击鼓传花的程序运行效果

8.4 阅读源代码

（1）Flower.java的代码如下：

```
import java.util.Random;                            //负责产生随机数的Random类
public class Flower {
     Random random;                                 //负责产生随机数
     Flower(){
         random = new Random();
     }
     public int getRamdomNumber(){
        int number = random.nextInt(8);    //返回0～7中的某个数
        return number+1;
     }
}
```

（2）People.java的代码如下：

```
public class People {
    int peopleNumber;                               //代号
    Flower flowerInhand;                            //手里拿的花
    People previousPeople;                          //传我花的人
    People nextPeople;                              //接我花的人
    public void setFlower(Flower flower){
        flowerInhand = flower;
        int n =flowerInhand.getRamdomNumber();
        if(n == 6)  {
           previousPeople.setNextPeople(nextPeople);
           nextPeople.setPreviousPeople(previousPeople);
           try{ Thread.sleep(1000);
           }
           catch(Exception exp){}
           System.out.println("代号"+peopleNumber+"的人退出击鼓传花");
        }
        if(this == nextPeople){
           System.out.println("代号"+peopleNumber+"是最后一个人");
           return;
        }
        nextPeople.setFlower(flowerInhand );    //传花给下一位
    }
    public void setPreviousPeople(People people){
        previousPeople = people;
    }
```

```
    public void setNextPeople(People people){
        nextPeople = people;
    }
}
```

（3）MainClass.java的代码如下：

```
public class MainClass {
    public static void main(String args[]) {
        Flower roseFlower = new Flower();
        int length = 12;
        if(length<2)
           return;
        People [] peopleInCircle = new People[length];
        for(int i=0;i<peopleInCircle.length;i++){
           peopleInCircle[i] = new People();
           peopleInCircle[i].peopleNumber =i+1;
        }
        for(int i=0;i<peopleInCircle.length;i++){
           if(i==0){
             peopleInCircle[i].setNextPeople(peopleInCircle[i+1]);
             peopleInCircle[i].setPreviousPeople(peopleInCircle[length-1]);
           }
           else if(i==length-1){
              peopleInCircle[i].setNextPeople(peopleInCircle[0]);
              peopleInCircle[i].setPreviousPeople(peopleInCircle[i-1]);
           }
           else {
              peopleInCircle[i].setNextPeople(peopleInCircle[i+1]);
              peopleInCircle[i].setPreviousPeople(peopleInCircle[i-1]);
           }
        }
        peopleInCircle[6].setFlower(roseFlower);
    }
}
```

请女朋友吃海鲜

9.1 场景故事

故事很简单，因为女朋友喜欢吃海鲜，男孩准备好海鲜和水果后，喂给女朋友吃，自己却一口也不吃。每次女朋友都会夸奖男孩的耐心和体贴。但是，如果男孩只准备了水果，没有准备海鲜，女朋友就会不高兴。男孩请女朋友吃海鲜如图9.1所示。

图9.1 男孩请女朋友吃海鲜

本故事纯属虚构，如有雷同，纯属巧合。

9.2 场景故事的目的

1. 侧重点

（1）两个类型相同的对象，一旦二者的引用相同，二者就具有完全相同的变量（实体）。

（2）对象用分配给自己的变量存储所需要的数据，调用类提供的方法操作数据。

（3）this是Java的一个关键字，可以出现在实例方法中，代表调用该实例方法的对象。

2．涉及的其他知识点

了解掌握一个设计模式——访问者模式。一个对象用分配给自己的变量存储所需要的数据。但在某些情况，可能不希望自己调用方法操作自己存储的数据，而是希望另一个对象来操作自己存储的数据，即在某些情况下，希望把数据的存储和处理解耦。比如，男孩准备了8只大虾（自己的数据），但希望女朋友处理大虾（吃大虾），而不是自己吃掉大虾（按照“习俗”，请女朋友吃饭，不能让女朋友带着大虾来）。男孩请女朋友吃海鲜这个例子虽然代码简单，但却蕴含着设计模式中的重要思想之一：把存储和处理解耦（访问者模式是把数据的存储和处理解耦的成熟模式之一）。

被访问者，如Boy的代表性代码如下：

```
public void inviteGirlfriend(Girl girl){
   girl.eatSeafood(this); //girl负责处理数据，体现数据处理和存储者解耦
}
```

访问者，例如Girl中的代表性代码是：

```
public void eatSeafood(Boy boy) {
   ……//操作boy的数据
}
```

3．进一步的尝试

在Girl类增加一个成员变量weight，当Boy类对象tonni请女孩anna吃饭时，代码如下：

```
tonni.inviteGirlfriend(anna)
```

首先判断anna的weight变量的值是多少，如果大于120就不执行girl.eatSeafood(this);。

9.3 程序运行效果与视频讲解

视频讲解

女朋友很爱吃海鲜，吃掉了男孩准备的大部分海鲜。主类是MainClass。程序运行效果如图9.2所示。

```
女孩吃海鲜和水果之前，男孩的海鲜和水果数量：
苹果:6      大虾:12     螃蟹:7

---女孩：很不错，下次海鲜品种可以再多拿来一些！---

女孩吃过后，男孩的海鲜和水果数量：
苹果:3      大虾:0      螃蟹:1
```

图9.2 请女朋友吃海鲜的程序运行效果

9.4 阅读源代码

（1）Boy.java的代码如下：

```
public class Boy {
   int apple;                       //男孩存储自己的苹果数量
   int prawn;                       //男孩存储自己的大虾数量
   int crab;                        //男孩存储自己的螃蟹数量
   boolean beingLiked;
   public void inviteGirlfriend(Girl girl){
       girl.eatSeafood(this); //负责处理数据，体现数据处理和存储者解耦
   }
}
```

（2）Girl.java的代码如下：

```
public class Girl {
    public void eatSeafood(Boy boy) {
        if(boy.prawn <= 0&&boy.crab<=0){
           System.out.println("你这是想请我吃海鲜吗？大骗子！");
           return;
        }
        if(boy.apple>=1) {
           boy.apple -= boy.apple/2;
        }
        if(boy.prawn>=1) {
```

```
            boy.prawn = 0;
        }
        boy.crab = boy.crab>1?1:0;
        boy.beingLiked = true;
        System.out.println("\n---女孩：很不错,下次海鲜品种再多拿来一些！---\n");
    }
}
```

（3）MainClass.java的代码如下：

```
public class MainClass {
    public static void main(String args[]) {
        Boy tonni = new Boy();
        tonni.apple = 6;
        tonni.prawn = 12;
        tonni.crab  = 7;
        Girl anna =new Girl();
        System.out.println("女孩吃海鲜和水果之前,男孩的海鲜和水果数量：");
        System.out.printf
          ("苹果:%-5d大虾:%-5d螃蟹:%-5d\n",tonni.apple,tonni.prawn,tonni.
        crab);
        tonni.inviteGirlfriend(anna);
        if(tonni.beingLiked){
            System.out.println("女孩吃过后,男孩的海鲜和水果数量：");
            System.out.printf
           ("苹果:%-5d大虾:%-5d螃蟹:%-5d\n",tonni.apple,tonni.prawn,tonni.
           crab);
        }
        else {
           System.out.println("女孩生气地走了");
           System.out.printf
           ("苹果:%-5d大虾:%-5d螃蟹:%-5d\n",tonni.apple,tonni.prawn,tonni.crab);
        }
    }
}
```

草船借箭

10.1 场景故事

草船借箭是我国古典名著《三国演义》中关于赤壁之战的一个故事。借箭由周瑜故意提出，他让诸葛亮在十日之内赶制出十万支箭。机智的诸葛亮一眼识破这是一条害人之计，却淡定地说道："曹操大军即日将至，若候十日，必误大事。"并表示："只需三天的时间，就可以办完复命。"后来，多亏了有鲁肃的帮忙，诸葛亮利用曹操多疑的性格，调用了几艘草船诱敌，终于借到了十万余支箭。

草船借箭故事中的草船上的草人是非常重要的角色，因为草人有接纳箭的能力（铁人就没有），而且需要特别注意故事的另外一个关键点：需要曹兵帮忙把箭射到草人上。草船借箭的故事如图10.1所示。

图10.1 草船借箭的故事

10.2 场景故事的目的

1. 侧重点

如果参数是对象，改变参数对象的变量（实体），就会导致原实参对象的变量发生同样的变化。方法的所有参数都是“传值”的，即方法中参数变量的值是调用者指定的值的复制。关键字this代表一个对象，可以出现在实例方法中，代表调用该方法的对象。

2. 涉及的其他知识点

掌握一个设计思想：当一个对象需要和另一个对象交互时，借助访问者模式可以合理地分配算法。比如，对于草人和曹兵，曹兵射箭到草人，那么就应该由草人决定箭头是否可以射在草人上。当草人被射满箭之后，就会不允许曹兵再射箭到草人上。因此，经常借助设计模式中访问者模式的思想设计代码，即曹兵是访问者，被访问者是草人。例如，在曹兵的SoldierOfCao类里有这样的代码（访问草人grassman，即向草人射箭）：

```
shootAarrow(Grassman grassman);
```

而草人是被访问者，相应的Grassman类里会有和shootAarrow(Grassman grassman) 相呼应的代码：

```
acceptArrow(SoldierOfCao soldier)
```

然后，在这个方法里形成算法，比如草人身上多了一支箭，曹兵就少了一支箭等。

3. 进一步的尝试

让程序能输出最先被射满的船是哪艘？最后被射满的是哪艘？

10.3 程序运行效果与视频讲解

草船上的某个草人被弓箭射满之后，曹兵的弓箭手还可能继续向该草人射箭，但该草人不再增加箭支的计数（射向它的那支箭有可能掉船上了，也可能掉江里了），但程序继续提示说“xxx草人已经被射满箭了”。主类是MainClass。程序运行效果如图10.2所示。

视频讲解

```
孔明号船23号草人已经被射满箭了
诸葛号船33号草人已经被射满箭了
 鲁肃号船7号草人已经被射满箭了
孔明号船14号草人已经被射满箭了
鲁肃号船15号草人已经被射满箭了
 孔明号船6号草人已经被射满箭了
 鲁肃号船9号草人已经被射满箭了
诸葛号船27号草人已经被射满箭了
诸葛号船17号草人已经被射满箭了
 诸葛号船4号草人已经被射满箭了
鲁肃号船29号草人已经被射满箭了

草船成功借箭100000支
```

图10.2　草船借箭的程序运行效果

10.4　阅读源代码

（1）Grassman.java的代码如下：

```
public class  Grassman {
    public static final int MAX = 1000;          //草人最多能承载的箭的数量
    int arrowAmount;                             //存放草人能承载的箭的数量
    String name;                                 //存放草人的名字
    public void acceptArrow(SoldierOfCao soldier) {
                                                 //有接收箭的能力acceptArrow
        if(arrowAmount<=MAX){
           arrowAmount++;
           soldier.arrows--;
        }
        else {
           soldier.arrows--;
           System.out.printf("%16s\n",name+"草人已经被射满箭了");
        }
    }
    public int getArrowAmount(){
        return arrowAmount;
    }
    public void setName(String s){
        name = s;
    }
}
```

（2）SoldierOfCao.java的代码如下：

```
public class SoldierOfCao {
```

```
    int arrows;                              //存放曹兵携带的箭的数量
    int number;                              //曹兵的代号
    public void shootAarrow(Grassman grassman){
        if(arrows>1){
           grassman.acceptArrow(this);
        }
        else if(arrows==1){
           grassman.acceptArrow(this);
           System.out.printf("%16s\n",number+"号曹兵把箭都射光了");
        }
    }
    public void setArrows(int n){
       if(n>=0)
         arrows = n;
    }
}
```

（3）GrassShip.java的代码如下：

```
public class  GrassShip {
    Grassman [] grassMan;
    GrassShip(String name){
       grassMan = new Grassman[35];
       for(int i=0;i<grassMan.length;i++){
          grassMan[i] = new Grassman();
          grassMan[i].setName(name+"船"+i+"号");
       }
    }
    public int getArrows()  {      //返回草船上全部草人身上的箭的数量之和
       int sum = 0;
       for(int i=0;i<grassMan.length;i++){
          sum += grassMan[i].getArrowAmount();
       }
       return sum;
    }
    public int getGrassmanAmount(){ //返回草人的数量
       return grassMan.length;
    }
}
```

（4）MainClass.java的代码如下：

```
import java.util.Random;
public class MainClass {
```

```
    public static void main(String args[]){
        Random random = new Random();
        SoldierOfCao [] soldier = new SoldierOfCao[5000];
        for(int i=0;i<soldier.length;i++) {
            soldier[i] = new SoldierOfCao();
            int arrows = random.nextInt(100)+1; //曹兵携带的箭的数目是随机的
            soldier[i].setArrows(arrows);
            soldier[i].number = i+1;
        }
        GrassShip grassShip1 = new GrassShip("诸葛号");
        GrassShip grassShip2 = new GrassShip("孔明号");
        GrassShip grassShip3 = new GrassShip("鲁肃号");
        int borrowDays = 10;                          //限10天造箭10万支箭
        int borrowArrows=100000;
        boolean isSuccess = true;
        while(grassShip1.getArrows()+
              grassShip2.getArrows()+
              grassShip3.getArrows() < borrowArrows) {
            borrowDays--;
            if(borrowDays==0){
                System.out.println("\n草船借箭失败");
                isSuccess = false;
                break;
            }
            for(int i=0;i<soldier.length;i++) {
                int index = random.nextInt(grassShip1.getGrassmanAmount());
                if(soldier[i].arrows>=1){
                    soldier[i].shootAarrow(grassShip1.grassMan[index]);
                }
                index = random.nextInt(grassShip2.getGrassmanAmount());
                if(soldier[i].arrows>=1){
                    soldier[i].shootAarrow(grassShip2.grassMan[index]);
                }
                index = random.nextInt(grassShip3.getGrassmanAmount());
                if(soldier[i].arrows>=1){
                    soldier[i].shootAarrow(grassShip3.grassMan[index]);
                }
            }
        }
        if(isSuccess)
            System.out.println("\n草船成功借箭"+borrowArrows+"支");
    }
}
```

男孩求婚

11.1 场景故事

很久以前，一个男孩因一见钟情而喜欢上一个女孩。于是，他决定来到女孩的家里求婚。

女孩的父亲问男孩："你知道我女儿的名字吗？"

男孩尴尬地摸摸头，回答道："我不知道她的名字。"

父亲说："好吧，我只有这一个宝贝女儿。现在，我可以告诉你线索，需要你自己去努力找出她的名字。她的名字是由小写的英文字母构成，我用'魔法'把女儿的名字中的每个英文字母都分别写在一张卡片上，但你看不到卡片上的英文字母。前方的花园里有26位仙女，每位仙女都代表一个小写的英文字母，并且每位仙女都能看见卡片上的英文字母。每次我会给你一张卡片，然后你可以进入花园并询问哪位仙女代表的字母与你手中卡片上的英文字母一致。如果一致，则那位仙女就会点头示意，并告诉你卡片上是哪个英文字母，否则她就会微笑地摇摇头，但会告诉你，你手中卡片上的英文字母是否比她代表的英文字母大或小。"男孩求婚如图11.1所示。

图11.1　男孩求婚

男孩临走前，父亲嘱咐道："切记不能把卡片同时给多个仙女看，但你可以分多次，每次只给一位仙女看卡片。当你走出花园时，要告诉我你一共给几位仙女看过卡片。如果你第一次进花园，运气非常好的话，只给一位仙女看了卡片，就可以知道卡片上的字母是哪个字母，那你肯定不会出局。如果我发现你某次进花园出来后，询问的仙女总数超出了我规定的数目，我就不会再给你卡片了，你就出局了。如果你最后知道了名字，就大声说出I love '名字'，你就可以娶她了！"

父亲见男孩一筹莫展的样子，补充道："我给你一个法宝吧！这个法宝是sort，只要你说出sort口令，那么这26位仙女会按英文字母由小到大重新列队。"

男孩十分感激地对这位父亲说："好吧，谢谢您！我一定能找出她的名字！"

男孩去求助名字是Arrays的智者，希望Arrays智者能给他一个锦囊办法。

Arrays智者对男孩说："锦囊办法我倒是有一个，那就是你可以直接使用我的名字。"

男孩不解地问道："我该怎样使用呢？"

Arrays智者给出了下面的锦囊办法。

如果数组c是排序的数组，想询问字符t是否在数组c里，代码如下：

```
int n = Arrays.binarySearch(c, 't' );
```

锦囊会告诉你一个整数n，即字符t在数组c中的位置。如果整数n是0，则表示字符t在数组c中排在第一个的位置；如果整数n是1，则表示字符t在数组c中排在第二个的位置；如果整数n小于0，则表示字符t不在数组c中。但我的锦囊无法告诉你每次问了几位仙女，你必须要自己想办法记住你已询问过几位仙女。

男孩受到Arrays智者的启发，认为智者的锦囊办法可以减少询问的次数。于是，他想这样利用这个锦囊办法。

第一步，将长度是为n的数组c排序。

第二步，把要寻找的字符x与c[n/2]进行比较。

如果x=c[n/2]，则找到字符x，算法中止；

如果x<c[n/2]，则只要在数组c的左半部分继续搜索字符x；

如果x>c[n/2]，则只要在数组c的右半部继续搜索字符x。

在数组c的左半部分或右半部分再重复这样的办法，如此下去就可以断定是否能搜索到x。

时间复杂度是while循环语句循环的循环次数。如果共有n个元素，那么while

循环语句每次查找的元素的范围（元素的个数）依次折半递减，分别是n，$n/2$，$n/4$，…，$n/2^k$。其中k就是循环的次数。最后让$n/2^k$取整后等于1，然后求解出的k的值，就是搜索所用的最多次数。令

$$n/2^k=1$$

可得

$$k=\log_2{}^n$$

因此，时间复杂度可以表示为$O(h)=O(\log_2{}^n)$，换底后就是$O(\log_{10}{}^n/\log_{10}{}^2)$。一共26个仙女，男孩单次询问过的仙女数不能超过：

$$\log_{10}{}^{26}/\log_{10}{}^2+1$$

这个值取整（舍弃小数部分）就是5，意味着如果男孩询问仙女数目超过了5，就会被赶出局。

男孩终于想出了自己的办法，达到了每次进入花园询问的仙女的总人数都没有超出女孩父亲的要求，成功地知道了女孩的名字。

后来，男孩和女孩幸福地走进了婚姻的殿堂，第二年有了一对可爱的儿女，哥哥起名叫sort（排序），妹妹起名叫binarySearch（二分查找）。

本故事纯属虚构，如有雷同，纯属巧合。

11.2 场景故事的目的

1. 侧重点

（1）巩固static方法，即静态方法的用法和思想。

静态方法可以用类名直接调用，如果一个算法不需要创建对象，就能很好地实现，那么可以将这样的方法定义为static方法。

（2）掌握Arrays类的static void sort(…)方法和static int binarySearch(…)方法。

2. 涉及的其他知识点

了解Math类中的常用static方法（即静态方法），如static double og10(double x)方法等。

3. 进一步的尝试

修改程序，让女孩的名字中可以有大写的英文字母。另外，你可以再给男孩提供一个找字母的锦囊方法，看看该锦囊方法的成功概率是多少？

11.3 程序运行效果与视频讲解

视频讲解

男孩终于知道了女孩的名字。主类是MainClass。男孩求婚的程序运行效果如图11.2所示。

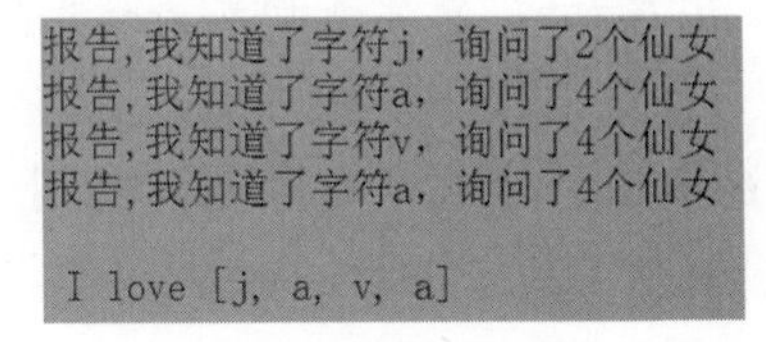

报告,我知道了字符j，询问了2个仙女
报告,我知道了字符a，询问了4个仙女
报告,我知道了字符v，询问了4个仙女
报告,我知道了字符a，询问了4个仙女
I love [j, a, v, a]

图11.2 男孩求婚的程序运行效果

11.4 阅读源代码

（1）FindCharOfName.java的代码如下：

```
//Arrays类的static方法binarySearch无法告诉你查找的次数
import java.util.Arrays;
public class FindCharOfName {
    static int count;                                //次数
    public static char getChar(char []c,char ch){
                                                     //在数组c中找字符ch
       Arrays.sort(c);                               //把组数c排序
       int indexStart =0;
       int indexEnd =c.length;
       int indexMiddle = (indexStart+indexEnd)/2;
       int d =(int)(Math.log10(c.length)/Math.log10(2))+1;
       count = 0;
       while(c[indexMiddle] != ch) {
          count++;
          if(count>=d){
             System.out.println("\n"+count+"次没找到");
             break;
```

```
            }
            if(ch<c[indexMiddle]){
               indexEnd = indexMiddle;
               indexMiddle = (indexStart+indexEnd)/2;
            }
            else {
               indexStart = indexMiddle;
               indexMiddle = (indexStart+indexEnd)/2;
            }
         }
         if(count>d) {
            return '*';
         }
         return c[indexMiddle];
     }

}
```

（2）MainClass.java的代码如下：

```
import java.util.ArrayList;
import java.util.Collections;
import java.util.Arrays;
public class MainClass {
   public static void main(String args[]) {
         char [] girlName ={'j','A','v','a'};
         char [] findName = {'*','*','*','*'};
         ArrayList<Character> list =
         new ArrayList<Character>();        //存放字符的数组表
         for(char letters='a';letters<='z';letters++) {
                                             //26个英文字母放入数组表
            list.add(letters);
         }
         Collections.shuffle(list);         //打乱数组表，即洗牌
         char [] c = new char[26];
         int d =(int)(Math.log10(c.length)/Math.log10(2))+1;
         for(int i=0;i<c.length;i++){
            c[i] = list.get(i);
         }
         for(int i=0;i<girlName.length;i++){
           char ok = FindCharOfName.getChar(c,girlName[i]);
           findName[i]= ok;
           if(FindCharOfName.count>=d){
              System.out.println("对不起你出局了");
```

```
                break;
            }
            System.out.println
            ("报告,我知道了字符"+""+ok+", 询问了"+FindCharOfName.count+"个
            仙女");
        }
        System.out.println("\n I love "+Arrays.toString(findName));
    }
}
```

接力赛跑

12.1 场景故事

4×100米接力中，运动员必须在长度为20米的接力区内完成交接棒，即所有接棒者均不可在接力区外起跑。接力棒必须拿在手上，直到交接棒或比赛结束为止。

在2012年伦敦奥运会的男子4×100米接力赛中，由尤塞恩·博尔特（Usain Bolt）领军的牙买加队以36秒84蝉联冠军，并且再一次刷新了世界纪录。而在2012年伦敦奥运会女子4×100米接力赛中，美国队以40秒82打破了沉睡27年的世界纪录，一举夺冠。目前，女子4×100米接力赛的世界纪录仍为40秒82。

注意，对于一个队中的4名接力队员，他们将共享400米距离的跑道，从第一棒至第四棒的队员，需要各自完成大约100米的有效计时距离（因为有接力区），最后总距离是400米。4×100米接力，如图12.1所示。

图12.1　4×100米接力

12.2 场景故事的目的

1. 侧重点

static关键字。如果类中有static变量，那么分配给该类的对象的这个static变量占有相同的一处内存，改变其中一个对象的static变量会影响其他对象的static变量，也就是说类的对象共享static变量。static变量是与类相关联的变量，即static变量是和该类创建的所有对象相关联的变量，改变其中一个对象的static变量就同时改变了其他对象的static变量。因此，static变量不仅可以通过某个对象访问，还可以直接通过类名访问。

2. 涉及的其他知识点

使用java.util.Random类的对象模拟接力区，使得运动员在交接棒时有犯规的可能。

3. 进一步的尝试

让程序分别输出四棒运动员所跑的有效距离。

12.3 程序运行效果与视频讲解

视频讲解

接力比赛的交接棒是很重要的环节之一，即使在重大的比赛中，交接棒犯规情况也时常发生。4×100米接力的程序运行效果图如图12.2所示。图12.2（a）表示交接棒没有犯规，图12.2（b）（主类是MainClass）显示了第3棒和第4棒的交接棒不在接力区内（接力区是290–310）。而二者的实际交接棒位置是在311米处。

```
第1棒负责跑到101米位置,犯规了吗？false
第2棒负责跑到200米位置,犯规了吗？false
第3棒负责跑到297米位置,犯规了吗？false
第4棒负责跑到400米位置,犯规了吗？false
```

(a) 没有犯规

```
第1棒负责跑到108米位置,犯规了吗？false
第2棒负责跑到195米位置,犯规了吗？false
第3棒负责跑到311米位置,犯规了吗？true
第4棒负责跑到400米位置,犯规了吗？false
```

(b) 出现犯规

图12.2 接力赛跑的程序运行效果

注意，整个比赛中，只要一人犯规成绩就作废，所以程序只需考虑一个人的犯规情况即可。

12.4 阅读源代码

（1）Relayathlete.java的代码如下：

```
import java.util.Random;
public class Relayathlete {                              //接力运动员
    public static int distence;                          //被接力运动员共享
    public boolean isFoul = false;                       //接力是否犯规
    Relayathlete nextPerson;                             //当前运动员的下一棒
    int nextPosition;                                    //下一个接力区的起点位置
    int number;                                          //存放棒号
    Relayathlete(int n) {
       number = n;
    }
    public void setNextPerson(Relayathlete person) {
        nextPerson = person;
    }
    public void setNextPosition(int position){           //设置下一个接力区的位置
        nextPosition = position;
    }
    public void run(){
       Random random = new Random();
       int realHandover =
       (nextPosition-1)+random.nextInt(23);              //实际交棒位置，这里故意
                                                         //设置有误差
       while(true){
           if(distence == 400){
              break;
           }
           distence = distence+1;
           if(distence == realHandover){
             if(realHandover<nextPosition||realHandover>nextPosition+20)
             {
                 isFoul = true;
             }
             break;
           }
       }
       System.out.printf("\n第%d棒负责跑到%d米位置,犯规了吗? %b",
                            number,distence,isFoul);
       if(nextPerson!=null)
          nextPerson.run();                              //下一棒开跑
    }
}
```

（2）MainClass.java的代码如下：

```
public class MainClass {
```

```
    public static void main(String args[]) {
        Relayathlete first = new Relayathlete(1);
        Relayathlete second = new Relayathlete(2);
        Relayathlete third = new Relayathlete(3);
        Relayathlete fourth = new Relayathlete(4);
        first.setNextPerson(second);
        first.setNextPosition(90);
        second.setNextPerson(third);
        second.setNextPosition(190);
        third.setNextPerson(fourth);
        third.setNextPosition(290);
        fourth.setNextPosition(0);
        first.run();
    }
}
```

高考录取分数线

13.1 场景故事

张明高考的那年，他所在的省的高考报考规则是在公布成绩之前填报志愿。张明依据往年的一本录取分数线的情况，认为自己可以报考一所重点大学。

张明的父母参加了一个有许多国内著名的大学参加的高考志愿填报咨询会。张明的妈妈经常问这些大学招生办的工作人员："请问，贵校的录取分数线大概是多少？"而得到的答复往往是根据近几年我校的录取数据分析，最低录取分数线通常比贵省的一本录取分数线约高出xx分等这样的回话。其中，也有一个热心的人举例解释说："我校近三年在贵省的最低录取分数线分别比贵省同年的一本录取分数线高71分、68分、72分，建议您今年可以参考75分，相对来说比较有把握一些。"

张明的妈妈发现一个现象，这些大学招生办的工作人员所说的"录取分数线"需要参考同年该省的一本录取分数线。高考志愿填报咨询，如图13.1所示。

图13.1 高考志愿填报咨询

13.2 场景故事的目的

1. 侧重点

（1）super关键字。

“高考录取分数线”咨询与Java语言中继承的某些情景类似，即子类需要重写父类的实例方法（如“录取分数线”），以便修改方法体的具体内容，但又需要父类的方法的行为或返回值（需要同年该省的一本录取分数线）。子类一旦重写了可以继承的方法，就覆盖了所继承的方法，那么子类创建的对象就不能调用被覆盖的方法，而关键字super可以调用被覆盖的方法。因此，如果在子类中想使用被子类覆盖方法就需要使用关键字super。

（2）上转型对象。上转型对象操作子类继承的方法或子类重写的实例方法，其作用等价于用子类对象去调用这些方法。因此，如果子类重写了父类的某个实例方法后，当对象的上转型对象调用这个实例方法时，一定是调用了子类重写的实例方法。

2. 涉及的其他知识点

private成员不能被子类继承，但可以调用继承的方法访问父类的private成员。

3. 进一步的尝试

在University类里增加一个public int getAdmissionLine()方法的重载方法public int getAdmissionLine(int x)方法。在该方法中直接返回public int getAdmissionLine()方法的返回值。子类重写public int getAdmissionLine(int x)方法时，可以利用参数x确定自己的预估分数比“同年该省的一本录取分数线”高了多少分。

13.3 程序运行效果与视频讲解

视频讲解

程序运行显示了XY大学和AAA大学的最低录取分数线，主类是MainClass。程序运行效果如图13.2所示。

```
...省一本录取分数线529
AAA大学的录取分数线624
...省一本录取分数线529
XY大学的录取分数线609
...省一本录取分数线529
XY大学的录取分数线609
...省一本录取分数线529
AAA大学的录取分数线624
```

图13.2 XY大学和AAA大学的最低录取分数线

13.4 阅读源代码

（1）University.java的代码如下：

```
public class University {
    private static final int line = 529;
    public int getAdmissionLine(){
       System.out.println("...省一本录取分数线"+line);
       return line;
    }
    public int compareTo(University university){
       int m=
       this.getAdmissionLine()-university.getAdmissionLine();
       return m;
    }
}
```

（2）XYUniversity.java的代码如下：

```
public class XYUniversity extends University{
    public int getAdmissionLine(){
       int line = super.getAdmissionLine()+80;
       System.out.printf("XY大学的录取分数线%d\n",line);
       return line;
    }
}
```

（3）AAAUniversity.java的代码如下：

```
public class AAAUniversity extends University {
```

```
    public int getAdmissionLine(){
       int line = super.getAdmissionLine()+95;
       System.out.printf("AAA大学的录取分数线%d\n",line);
       return line;
    }
}
```

（4）MainClass.java的代码如下：

```
public class MainClass {
    public static void main(String args[]) {
        University university = new AAAUniversity();
                                          //university是上转型对象
        int aaa = university.getAdmissionLine();
        University university2 = new XYUniversity();
                                          //university2是上转型对象
        int xy = university2.getAdmissionLine();
        int sub = university2.compareTo(university);
        System.out.printf("\n两所大学录取分数线相差%d",Math.abs(sub));
    }
}
```

一骑红尘妃子笑

14.1 场景故事

荔枝在唐朝时期，有“百果之中无一比”和“百果之王”的美誉。由于杨贵妃非常喜欢吃荔枝，特别是新鲜的、刚采摘的荔枝。对杨贵妃宠爱有加的唐玄宗想尽办法为她寻觅新鲜的，只为博杨贵妃的红颜一笑。

于是，唐玄宗派人将荔枝从盛产荔枝的岭南运至长安，途经2100余公里的路程，每隔一段距离就设有驿站。驿卒们都在各自的驿站等候着，每当负责运送荔枝的马匹接近驿站时，就将早已备好的马匹牵出来，驿卒接过装有荔枝的竹筒，立刻扬鞭飞驰而去，奔向下一个驿站。运送荔枝图，如图14.1所示。经过千难万险，用千里接力的方式运送的荔枝仿佛刚从树上采摘一般。唐代杰出的诗人杜牧写下流传至今的诗篇。

长安回望绣成堆，山顶千门次第开。
一骑红尘妃子笑，无人知是荔枝来。

图14.1　运送荔枝图

14.2 场景故事的目的

1. 侧重点

本场景的目的是巩固面向对象中一个基础核心思想——面向抽象设计。场景故事里的关键点是运送荔枝的行动一旦启动，在达到目的地之前就不能间断。从软件设计角度来看，荔枝对象应该组合马对象，以便更换所组合的对象，这有利于软件的维护（注意，马组合荔枝，不小心容易换错荔枝），即荔枝是更重要的对象，马是为之服务的重要组成部分。只有这样，才能模拟出“骏马交替千里行，只为一枝鲜荔枝”的效果，即软件一旦启动，在不终止软件运行的前提下，给软件动态地添加新的模块，并最终完成软件要达到的目的。

2. 涉及的其他知识点

Java反射。Class是java.lang包中的类，Class的实例用于封装和类有关的信息（即类型信息）。任何类都默认有一个public的静态的Class对象，该对象的名字是class，该对象封装当前类的有关信息（即类型的信息），如该类有哪些构造方法、成员变量和方法等。也可以让类的对象调用getClass()方法（从java.lang.Object类继承的方法）返回这个Class对象：class。

得到一个类的实例的最常用的方式就是使用new运算符和类的构造方法。Java也可以使用和类相关的Class对象得到一个类的实例，可以用Class类的下列static方法先得到一个和className类相关的Class对象：

```
public static Class<?> forName(String className) throws ClassNotFoundException
```

即方法返回一个和参数className指定的类相关的Class对象。再让这个Class对象调用

```
public Constructor<?> getDeclaredConstructor() throws SecurityException
```

方法得到className类的无参数的构造方法（因此className类必须保证有无参数的构造方法）。然后，Constructor<?>对象，调用newInstance()方法返回一个className类的对象。从JDK9版本开始，Class类的newInstance()方法被宣布为Deprecated（已过时）。

Class<?>，Constructor<?>和泛型知识有关。Class<?>中的?是个统配泛型，即?可以代表任何类型（不包括基本类型）。

3. 进一步的尝试

让程序输出每匹马跑过的距离。

14.3 程序运行效果与视频讲解

视频讲解

程序运行效果如图14.2所示。主类是MainClass。在不终止已运行的程序的前提下，继续编辑、编译Horse类的子类，模拟即将要更换的马，即用Horse类的子类创建对象。比如，想建立一个名字是RedHorse的子类，并用该子类创建对象，那么首先将RedHorse的子类保存到程序规定的目录中，然后编译通过（需要另行打开一个命令行窗口），那么正在运行中的程序就可以使用RedHorse的子类创建对象了。

```
输入要更换的马(回车不换):

到达                    1820km处
千里枣红马
输入要更换的马(回车不换):
WhiteHorse
到达                    2020km处
白龙马
输入要更换的马(回车不换):
BlackHorse
到达                    2080km处
白蹄黑色
输入要更换的马(回车不换):

到达                    2100km处
一骑红尘妃子笑，无人知是荔枝来
```

图14.2 骏马交替千里行

14.4 阅读源代码

（1）Horse.java的代码如下：

```
public abstract class Horse {
   public abstract String giveMess();
   public abstract int runDistance(int upperLimit);
}
```

（2）BlackHorse.java的代码如下：

```
public class BlackHorse extends Horse {
    int count = 0;                                    //每次稍作休息后，跑的次数
    public String giveMess() {
        return "白蹄黑色";
    }
    public int runDistance(int upperLimit){
        count++;
        if(count == 1) {                              //首次跑60km
           return
           60<upperLimit?60:upperLimit;
        }
        else if(count == 2) {                         //第2次能跑50km
           return
           50<upperLimit?50:upperLimit;
        }
        else if(count == 3) {                         //第3次能跑10km
           return
           10<upperLimit?10:upperLimit;
        }
        else {
           return 0;                                  //不能再跑了
        }
    }
}
```

（3）WhiteHorse.java的代码如下：

```
public class WhiteHorse extends Horse {
    int count = 0;                                    //每次稍作休息后，跑的次数
    public String giveMess() {
        return "白龙马";
    }
    public int runDistance(int upperLimit){
         count++;
        if(count == 1) {                              //首次跑200km
           return
           200<upperLimit?200:upperLimit;
```

```
        }
        else if(count == 2) {                            //第2次能跑180km
            return
            180<upperLimit?180:upperLimit;
        }
        else if(count == 3) {                            //第3次能跑150km
            return
            150<upperLimit?150:upperLimit;
        }
        else if(count == 4) {                            //第4次能跑70km
            return
            70<upperLimit?70:upperLimit;
        }
        else {
            return 0;                                    //不能再跑了
        }
    }
}
```

（4）RedHorse.java的代码如下：

```
public class RedHorse extends Horse {
    int count = 0;                                       //每次稍作休息后，跑的次数
    public String giveMess() {
        return "千里枣红马";
    }
    public int runDistance(int upperLimit){
        count++;
        if(count == 1) {                                 //首次跑800km
            return
            800<upperLimit?800:upperLimit;
        }
        else if(count == 2) {                            //第2次能跑500km
            return
            500<upperLimit?500:upperLimit;
        }
        else if(count == 3) {                            //第3次能跑350km
            return
            350<upperLimit?350:upperLimit;
        }
        else if(count == 4) {                            //第4次能跑170km
            return
            170<upperLimit?170:upperLimit;
        }
```

```
        else {
            return 0;                                  //不能再跑了
        }
    }
}
```

（5）MainClass.java的代码如下：

```
import java.util.Scanner;
public class MainClass {
    public static void main(String arg[]) {
        int distance = 2100;                           //模拟2100km
        int sum = 0;
        Litchi lychee =new Litchi();
        Horse horse = null;
        while(sum < distance) {
            try{
              lychee.currentState();
              System.out.println("输入要更换的马(回车不换):");
              Scanner scanner = new Scanner(System.in);
              String name = scanner.nextLine();
              Class<?> cs = Class.forName(name);
              horse=(Horse)cs.getDeclaredConstructor().newInstance();
             //如果没更换，则跳到catch
              lychee.setHorse(horse);                  //更换马匹
              sum += horse.runDistance(distance-sum);
              System.out.printf("到达%20s km处",sum);
            }
            catch(NoClassDefFoundError exp){
                if(horse!=null){
                  sum += horse.runDistance(distance-sum);
                  System.out.printf("到达%20s km处",sum);
                }
            }
            catch(Exception exp){
                if(horse!=null){
                   sum += horse.runDistance(distance-sum);
                   System.out.printf("到达%20s km处",sum);
                }
            }
        }
        System.out.printf("\n%-50s","一骑红尘妃子笑,无人知是荔枝来");
    }
}
```

场景故事 15

画龙点睛与画蛇添足

15.1 场景故事

传传说古时候有个画家叫张僧繇，他画龙画得特别好。一次，他在金陵（现南京市）安乐寺的墙壁上画了4条活灵活现、形象逼真的巨龙，只是这些巨龙都没有眼睛。人们问张僧繇："为什么不把眼睛画出来。"他说："眼睛可不能轻易画呀！一旦画了，龙就会腾空飞走的！"大家听后，谁也不相信他的话。后来，在人们多次请求下，张僧繇只好答应把龙的眼睛画出来。奇怪的事情果然发生了，当他刚刚点出第二条龙的眼睛时，突然刮起了大风，顷刻间电闪雷鸣，两条巨龙转动着光芒四射的眼睛冲天而起，腾空而去。成语"画龙点睛"就是从这个传说中来的。现在一般用来比喻做完某件事情后，在关键性的地方改动或增加富有创新特色的事物。画龙点睛如图15.1所示。

图15.1 画龙点睛

古时候，楚国有一家人，在祭完祖之后，准备将祭祀用的一壶酒，赏给帮忙办事的人喝。帮忙办事的人有很多，如果大家都喝这壶酒是不够的。若只是给一个人喝，那他能喝得有余。那么，这一壶酒到底该怎么分呢？

大家都安静下来，这时有人建议：每个人在地上画一条蛇，谁画得快，这壶酒就归谁喝。大家都认为这是一个好方法，便同意这样做。于是，大家在地上画起蛇来。

有个人画得很快，最先画好蛇，便端起酒壶准备喝酒。但是他回头看看其他人，都还没有画好呢。他心里想：他们画得真慢！他洋洋得意地说："你们画得好慢啊！我再给蛇画几只脚也不算晚呢！"于是，他便左手提着酒壶，右手给蛇画起脚来。

正在他一边给蛇画脚，一边说话的时候，另外一个人已经画好了。那个人马上把酒壶从他手里夺过去，说："你见过蛇吗？蛇是没有脚的，你为什么要给它添上脚呢？所以第一个画好蛇的人不是你，而是我了！"那个人说罢就仰起头来，咕咚咕咚地把酒喝下去了。

画蛇添足后比喻做了多余的事，非但无益，反而不合适；也比喻虚构事实，无中生有。画蛇添足如图15.2所示。

图15.2 画蛇添足

15.2 场景故事的目的

1. 侧重点

上面的场景故事里的蕴含着软件设计的一个重要思想，即所谓的"增强对象的能力"设计模式中的"装饰模式"。"装饰模式"是增强对象能力的经典的模式之一（从软件设计的逻辑角度来看，画龙点睛和给蛇添足没有区别）。如果在软件设计中实现"增强对象的能力"得到用户的好评，用户可能好评："画龙点睛"；否则差评："给蛇添足"。所以，设计模式不可乱用，应以满足用户需求作为根本出发点。

装饰模式属于结构型模式，结构中通常包括3种角色。

（1）抽象组件（Component）：抽象组件（抽象类）定义了需要进行装饰的方法。抽象组件就是“被装饰者”角色。例如，ImportantWord类。

（2）具体组件：具体组件是抽象组件的一个子类。例如，TimeWord类。

（3）装饰（Decorator）：该角色是抽象组件的一个子类，是“装饰者”角色，其作用是装饰具体组件（装饰“被装饰者”），因此“装饰者”需要包含“被装饰者”的引用。“装饰者”既可以是抽象类，又可以是一个非抽象类。例如，HuaLongDianjing类和HuaSheTianzu类。

2. 涉及的其他知识点

一个设计模式（Pattern）是针对某一类问题的最佳解决方案，而且已经被成功地应用于许多系统的设计中。它能够解决某种特定情景中重复发生的某个问题，即一个设计模式是从许多优秀的软件系统中总结出的、成功的、可复用的设计方案。

框架不是设计模式，而是针对某个领域，提供用于开发应用系统的类的集合。程序设计者可以使用框架提供的类来设计一个应用程序，而且在设计应用程序时，可以针对特定的问题使用某个设计模式。

3. 进一步的尝试

在ImportantWord类中增加一个 public abstract double getSum(int n);方法。

利用TimeWord类重写该方法，返回1~n的连续代数和。

利用HuaLongDianjing类的public double getSum(int n)方法，输出1~n的连续代数和，并返回1~n的连续代数和，再除以n的结果。

利用HuaSheTianzu类的public double getSum(int n)方法，输出1~n的连续代数和，并返回1~n的连续代数和，再乘以n的结果。

15.3 程序运行效果与视频讲解

视频讲解

程序中相当于“龙”或“蛇”的类是TimeWord类，可以给出英文单词，相当于画龙点睛的类是HuaLongDianjing类，可以给出英文单词并跟随汉语解释。主类是MainClass。程序运行效果如图15.3所示。

```
past
now
future
------- --------
past:过去
now:现在
future:将来
------- --------
past:These things happened in the past years.
now:She's a student now
future:In the future we need more Java programmers
------- --------
past:过去:These things happened in the past years.
now:现在:She's a student now
future:将来:In the future we need more Java programmers
------- --------
```

图15.3 增强对象的能力

15.4 阅读源代码

（1）ImportantWord.java的代码如下：

```
public abstract class ImportantWord{
   public abstract String[] getImportantWord();
}
```

（2）TimeWord.java的代码如下：

```
public class TimeWord extends ImportantWord{
    public String[] getImportantWord(){
       String [] wordArray = {"past","now","future"};
       return wordArray;
    }
}
```

（3）HuaLongDianjing.java的代码如下：

```
public class HuaLongDianjing extends ImportantWord{
    ImportantWord importantWord;
    String chinese[] ={"过去","现在","将来"};
    public HuaLongDianjing(ImportantWord importantWord){
      this.importantWord = importantWord;
    }
    public String[] getImportantWord(){
        String [] time = importantWord. getImportantWord();
        time = dianjing(time);
        return time;
```

```
    }
    private String[] dianjing(String []a){
        String [] str = new String[a.length];
        for(int i = 0;i<a.length;i++){
           str[i] = a[i]+":"+chinese[i];
        }
        return str;
    }
}
```

（4）HuaSheTianzu.java的代码如下：

```
public class HuaSheTianzu extends ImportantWord{
    ImportantWord importantWord;
    String sentence[] =
    {"These things happened in the past years.",
     "She's a student now",
     "In the future we need more Java programmers"};
    public HuaSheTianzu(ImportantWord importantWord){
      this.importantWord = importantWord;
    }
    public String[] getImportantWord(){
        String [] time = importantWord.getImportantWord();
        time = tianzu(time);
        return time;
    }
    private String[] tianzu(String []a){
        String [] str = new String[a.length];
        for(int i = 0;i<a.length;i++){
           str[i] = a[i]+":"+sentence[i];
        }
        return str;
    }
}
```

（5）MainClass.java的代码如下：

```
import java.util.Arrays;
public class MainClass{
   public static void main(String args[]){
      ImportantWord words = new TimeWord();
      String str[] = words.getImportantWord();
      for(String s:str) {
         System.out.println(s);
```

```
        }
        System.out.println("------- --------");
        ImportantWord wordsOne = new HuaLongDianjing(words);
        str = wordsOne.getImportantWord();
        for(String s:str) {
           System.out.println(s);
        }
        System.out.println("------- --------");
        ImportantWord wordsTwo = new HuaSheTianzu(words);
        str = wordsTwo.getImportantWord();
        for(String s:str) {
           System.out.println(s);
        }
        System.out.println("------- --------");
        ImportantWord wordsThree = new HuaSheTianzu(wordsOne);
        str = wordsThree.getImportantWord();
        for(String s:str) {
           System.out.println(s);
        }
        System.out.println("------- --------");
    }
}
```

非诚勿扰

16.1 场景故事

电视相亲节目是指由电视台策划制作的、为未婚男女提供相亲平台的电视节目。电视相亲节目的形式为邀请自愿报名参加节目的男女嘉宾来到录制现场，通过一定的环节，使两情相悦的男女嘉宾配对成功，从而达到相亲的目的。《非诚勿扰》是江苏卫视制作的一档生活服务类节目。该节目中有24位单身女生以灭灯或亮灯方式来决定报名男嘉宾在本场节目中的去留。《非诚勿扰》现场如图16.1所示。

图16.1 《非诚勿扰》现场

16.2 场景故事的目的

1. 侧重点

（1）面向接口编程的设计思想。

面向接口编程是指当设计某种重要的类时，不允许该类面向具体的类，而是面向接口，即所设计类中的重要数据是接口声明的变量，而不是具体类声明的对象。面向接口编程目的是为了应对用户需求的变化，将某个类中经常因需求变化而需要改动的代码从该类中分离出去。面向接口编程的核心是让类中每种可能的变化对应地交给实现接口的一个类去负责。

在《非诚勿扰》中，女性Female类，可能经常需要修改她的择偶标准，即修改analyzeData(Male man)方法体的内容，那么就将这种变化交给用于实现Filter接口的类去完成，而Female类只要面向Filter接口去设计代码即可。Female类如果想更改自己的择偶标准，只调用void setFilter(Filter filter)方法更换filter接口变量所引用的对象即可，而不需要修改其他任何代码。

（2）了解策略模式。

《非诚勿扰》中各个类的设计恰好符合策略模式。策略模式主要由下3部分组成。

① 策略（Strategy）：策略是一个接口，定义了若干个抽象方法。例如，Filter接口。

② 上下文（Context）：上下文是面向策略接口的类，即上下文包含有用策略声明的变量。例如，Female类。

③ 具体策略（Concrete Strategy）：具体策略是实现策略接口的类。例如，用Lambda表达式实现的Filter接口的匿名类。

策略模式满足“开–闭原则”，所谓“开–闭原则”（Open–Closed Principle）就是让你的设计对扩展开放，对修改关闭，即当一个设计中增加新的模块时，不须要修改现有的模块。在策略模式中，当增加新的具体策略时，无须修改上下文类的代码，就可以引用新的具体策略的实例。

（3）了解访问者模式。

一个对象用分配给自己的变量存储所需要的数据，但在某些情况，可能不希望自己调用方法对自己数据进行处理或判断，而是希望另一个对象来操作或判断自己的数据。例如，《非诚勿扰》中的男嘉宾（被访问者角色）负责提供自己相关的数据，女嘉宾（访问者角色）负责判断数据，并做出自己的决定：亮灯或灭灯。

2. 涉及的其他知识点

（1）Lambda表达式。Lambda表达式就是一个匿名方法（函数）。

① 通常的方法，如下所示：

```
int f(int a,int b ) {
    return a+b;
}
```

② Lambda表达式，如下所示：

```
(int a,int b ) -> {
   return a+b;
}
```

Lambda表达式就是只写参数列表和方法体的匿名方法（参数列表和方法体之间的符号是->）：

```
(参数列表) -> {
       方法体
 }
```

（2）与接口有关的Lambda表达式。

Java中使用Lambda表达式的主要目的是在使用单接口（只含有一个方法的接口）匿名类时，让代码更加简洁。例如：

```
interface Com {
    double computer(double x);
}
public class E {
   public static void main(String args[]) {
       Com com = new Com(){       //如果更改了接口里方法的名字，则匿名类代码受影响
        public double computer(double x){      //匿名类
                 return x*x*x;
                 } };
        System.out.println(com.computer(5));
        com =(double x)-> {       //Lambda表达式简化了匿名类的用法
               return x*x*x;  //如果更改了接口里方法的名字，则Lambada不受影响
               };
        System.out.println(com.computer(5));
   }
}
```

Lambda表达式过于简化，因此必须在特殊的上下文中，编译器才能推断出到底是哪个方法。因此Java中使用Lamada表达式的主要目的是让单接口（只含有一个方法）的匿名类的代码更加简洁和容易维护。

3．进一步的尝试

让程序模拟24位女嘉宾，并模拟灭灯者说的话。例如，不满意男性的哪些数

据（即输出一句话）。

16.3 程序运行效果与视频讲解

视频讲解

程序模拟有3位女嘉宾参与《非诚勿扰》，主类是MainClass。程序运行效果如图16.2所示。

```
第1号的灯的状态false
第2号的灯的状态true
第3号的灯的状态true
```

图16.2　有1位女嘉宾灭灯的运行效果

16.4 阅读源代码

（1）Filter.java的代码如下：

```
public interface Filter {
   public boolean analysis(Male male);
}
```

（2）Male.java的代码如下：

```
public class Male {
   int height;                                    //身高
   int weight;                                    //体重
   int annualSalary;                              //年薪
   public void provideDatae(Female female){
       female.analyzeData(this);                  //分析数据
   }
   public void setHeight(int height){
       this.height = height;
   }
   public void setWeight(int weight){
       this.weight = weight;
   }
   public void setAnnualSalary(int salary){
       annualSalary = salary;
```

```
    }
}
```

（3）Female.java的代码如下：

```
public class Female {
    boolean lightOn = true;                                    //亮灯状态
    Filter filter;
    public void analyzeData(Male man) {
       lightOn = filter.analysis(man);
    }
    public void setFilter(Filter filter){
       this.filter = filter;
    }
}
```

（4）MainClass.java的代码如下：

```
public class MainClass {
    public static void main(String args[]) {
       Male tonni = new Male();
       tonni.setHeight(178);
       tonni.setWeight(76);
       tonni.setAnnualSalary(120000);
       Female [] female = new Female[3];
       for(int i = 0;i<female.length;i++){
          female[i] = new Female();
       }
       female[0].setFilter((Male male)->{                      //Lambda表达式
          boolean boo = true;
          if(male.height<180) {
             boo = false;
          }
          if(male.weight>88){
             boo = false;
          }
          if(male.annualSalary<90000){
             boo = false;
          }
          return boo;
       });
       female[1].setFilter((Male male)->{                      //Lambda表达式
          boolean boo = true;
          if(male.height<170) {
```

```
            boo = false;
        }
        if(male.weight>90){
            boo = false;
        }
        if(male.annualSalary>=1000000){
            boo = true;
        }
        return boo;
    });
    female[2].setFilter((Male male)->{ //Lambda表达式
        boolean boo = true;
        if(male.annualSalary<70000) {
            boo = false;
        }
        if(male.weight>87){
            boo = false;
        }
        if(male.height>=190){
            boo = true;
        }
        return boo;
    });
    for(int i = 0;i<female.length;i++){
        tonni.provideDatae(female[i]);
    }
    for(int i = 0;i<female.length;i++){
        System.out.printf("\n第%d号的灯的状态%b",i+1,female[i].
        lightOn);
    }
  }
}
```

青山原不老，绿水本无忧

17.1 场景故事

有很多流传千古的古代诗句，有不少经典诗句都出自文人墨客的即兴发挥。

清人李文甫，少时随老师出游，老师指着远处覆盖着皑皑白雪的山峰，吟了一个上句："青山原不老，为雪白头。"要求学生对出下句。李文甫低头沉思，见身边一口池塘，绿柳倒映水中，清风吹拂，泛起阵阵涟漪，遂悟出一个对句："绿水本无忧，因风皱面。"风景如图17.1所示。

图17.1 青山原不老，为雪白头

唐伯虎同祝枝山因事到乡村，祝枝山看到家夫车水，给出上句："水车车水，水随车，车停水止。"唐伯虎即刻给出下句："风扇扇风，风出扇，扇动风生。""祝唐之对"实属巧妙，传诵至今。

某日朱元璋与刘伯温下棋。朱元璋给出上句："天作棋盘星作子，日月争光。"刘伯温对道："雷为战鼓电为旗，风云际合。"朱刘上、下句各合身份，用词堪称绝妙。

17.2 场景故事的目的

1. 侧重点

上面的场景故事里蕴含着软件设计的一个重要思想，即动态地扩展一个对象的功能（增强对象的能力），但不要改变原始类代码（这种思想被总结为设计模式中的装饰模式）。

在场景故事中，给出上句的人（相当于原始类代码角色），给出上句的人并不需要知道下句。而通过上句给出下句的人（相当于动态扩展角色），需要知道上句的内容，才能对出下句。

从面向对象的角度来看，任何动态扩展角色都需要原始类代码角色才能完成扩展任务。原始类代码角色和动态扩展角色应该具有相同的父类，即属于一类事情。

注意，经动态扩展角色最后形成的是一个完整的“对句”，只不过其中的下句恰好是动态角色的特色工作之一。

2. 涉及的其他知识点

装饰模式是23个经典设计模式中非常重要的模式之一，也是应用场合非常多的模式之一。也许你在许多设计中已经用过这个模式，只是不知道自己所使用的模式就是装饰模式。这也非常符合设计模式的来历，一个设计模式是针对某一类问题的最佳解决方案，而且已经被成功地应用于许多系统的设计中，它解决了在某种特定情景中重复发生的某个问题。因此，设计模式（Pattern）是从许多优秀的软件系统中总结出的、成功的、可复用的设计方案。Erich Gamma、Richard Helm、Ralph Johnson和John Vlissides这4人，经过4年多的艰苦努力和不懈工作，从众多优秀的软件中总结了23个设计模式，并于1994年著书出版。人们习惯地将这23个模式称作经典设计模式，将4位作者称为GOF，将他们的著作《Design Patterns：Elements of Reusable Object-Oriented Software》称为GOF之书。

3. 进一步的尝试

让NextSentence类中的giveSentence()方法除了返回对句，还能输出整个对句的长度。

17.3 程序运行效果与视频讲解

视频讲解

程序中FirsetSentce类的对象（相当于给出上句的人）可以给出对句中的上句。NextSentence类的对象（相当于利用上句、再给出下句的人）为动态角色。主类是MainClass。程序运行效果图如图17.2所示。

```
青山原不老，为雪白头。
绿水本无忧，因风皱面。
-----
水车车水，水随车，车停水止。
风扇扇风，风出扇，扇动风生。
-----
天作棋盘星作子，日月争光。
雷为战鼓电为旗，风云际合。
-----
```

图17.2 动态角色完成对句

17.4 阅读源代码

（1）Sentence.java的代码如下：

```
public abstract class Sentence{
   public abstract String giveSentence(int index);
   public void setSentence(Sentence sentence){
   }
}
```

（2）FirstSentence.java的代码如下：

```
public final class FirstSentence extends Sentence {
   String [] sentence;//存放上句
   int index = 0;
   FirstSentence(){
      sentence = new String[3];
      sentence[0] ="青山原不老，为雪白头.";
      sentence[1] ="水车车水，水随车，车停水止.";
      sentence[2] ="天作棋盘星作子，日月争光.";
   }
   public String giveSentence(int index){
      return sentence[index];
   }
}
```

（3）NextSentence.java的代码如下：

```
public class NextSentence extends Sentence {
    Sentence sentence;//需要存放一个上句
    public void setSentence(Sentence sentence){
       this.sentence = sentence;
    }
    public String giveSentence(int index){
       String str = sentence.giveSentence(index);
       if(index == 0){
          str = str+"\n绿水本无忧，因风皱面。";
       }
       else if(index == 1){
          str = str+"\n风扇扇风，风出扇，扇动风生。";
       }
       else if(index == 2){
          str = str+"\n雷为战鼓电为旗，风云际合。";
       }
       return str;
    }
}
```

（4）MainClass.java的代码如下：

```
public class MainClass{
   public static void main(String args[]){
      Sentence  firstSentence = new FirstSentence();
      Sentence nextSentence = new NextSentence();
      nextSentence.setSentence(firstSentence);
      String result = nextSentence.giveSentence(0);
      System.out.printf("\n%8s\n-----",result);
      result = nextSentence.giveSentence(1);
      System.out.printf("\n%8s\n-----",result);
      result = nextSentence.giveSentence(2);
      System.out.printf("\n%8s\n-----",result);
   }
}
```

三十六计走为上

18.1 场景故事

三十六计中有很多脍炙人口的计策，比如，第一计瞒天过海、 第二计围魏救赵、……第三十一计美人计、第三十二计空城计、第三十四计苦肉计、第三十六计走为上，如图18.1所示。

图18.1 第三十六计走为上

18.2 场景故事的目的

1. 侧重点

自定义异常和try-catch语句。在Java程序设计中需要对程序认为的异常进行必要的处理或提示，因为当这些异常发生后，可能会对程序的运行产生不利的影响或导致程序无法继续运行。即当try部分“陷于困境”时，立刻结束try部分，这就是所谓的“走为上”。执行catch部分，在catch部分提示用户或找到对程序有利的办法。

2. 涉及的其他知识点

注意，不要把自定义异常理解为程序出现了错误。因为，在某些应用程序里某种操作是合情合理的操作，而在另一个程序中可能就要对这样操作进行异常处理。

3. 进一步的尝试

假设客车的限载人数是66人，当客车运载人数超过66人后（如多了29人），会触发DangerException异常，在catch部分执行卸载，代码如下：

```
persons = max -n;                //persons的值是 -29
try {
   bus.loading(persons);        //下去persons乘客
}
catch(Exception exp) {
}
```

如果将其中的

```
persons = max -n;
```

更改为：

```
persons = max -n+1;              //-28
```

就意味着又会触发DangerException异常，但这次程序却没有对触发的DangerException异常给出具体处理的代码。请调试程序，观察Bus最后承载了多少人？

当在catch部分再次执行try-catch语句准备处理同样的异常时，通常处理方式是在其catch部分执行return语句强制结束当前方法的执行，或执行Sysytem.exit(0);结束程序的运行，即不允许再犯同样的错误，代码如下：

```
persons = max -n+1;
try {
    bus.loading(persons);       //下车的persons乘客
}
catch(Exception exp){
    return;
}
```

18.3 程序运行效果与视频讲解

营运客车超员是交通安全的隐患，应当进行必要的处理。当客车（限载是66

人）运载人数超出了29人后触发DangerException异常，处理异常时卸载29人。主类是MainClass。处理异常程序运行效果如图18.2所示。

视频讲解

```
目前乘坐了6位乘客
目前乘坐了18位乘客
目前乘坐了40位乘客
95人--超员!
进行超员处理.
目前乘坐了66位乘客
目前bus载客66人
```

图18.2 处理异常程序运行效果

18.4 阅读源代码

（1）DangerException.java的代码如下：

```
public class DangerException extends Exception {
   final String message = "--超员! ";
   int dangerNumber;
   DangerException(int n){
      dangerNumber = n;
   }
   public String warnMess() {
       return ""+dangerNumber+"人"+message;
   }
}
```

（2）Bus.java的代码如下：

```
public class Bus {
     int numberOfpeople;  //承载人数
     int maxNumber;        //最多可运载人数
     public void setMaxNumber(int number) {
         maxNumber = number;
     }
     public void loading(int m) throws DangerException {
        numberOfpeople += m;
        if(numberOfpeople>maxNumber) {
           throw new DangerException(numberOfpeople);
```

```
        }
        System.out.println("目前乘坐了"+numberOfpeople+"位乘客");
    }
    public int getNumberOfpeople(){
        return numberOfpeople;
    }
}
```

（3）MainClass.java的代码如下：

```
public class MainClass {
   public static void main(String args[]) {
      Bus bus = new Bus();
      int max = 66;
      bus.setMaxNumber(max);
      int persons =6;
      try{
            bus.loading(persons);
            persons = 12;
            bus.loading(persons);
            persons = 22;
            bus.loading(persons);
            persons = 55;
            bus.loading(persons);
      }
      catch(DangerException e) {
            System.out.println(e.warnMess());
            System.out.println("进行超员处理.");
            int n = bus.getNumberOfpeople();
            persons = max -n ;
            try {
               bus.loading(persons);      //有persons个乘客下车
            }
            catch(Exception exp){}
      }
      System.out.println("目前bus载客"+bus.getNumberOfpeople()+"人");
   }
}
```

零钱魔盒

19.1 场景故事

很久以前，在一个海边的小镇上，有一个Java杂货店。Java杂货店里就只有老奶奶和8岁的孙子皮皮卡，皮皮卡的父母经常出海打鱼。有一天，来了一位魔法师，想买用于削土豆皮的小刀。

皮皮卡一听说他是魔法师，连忙说道："因为奶奶和我每次找零钱都很慢，而且时常出错，您能给我们变出一个可以自动找零钱（找赎）的盒子吗？"

魔法师说："我现在没有魔力了，必须要吃很多土豆泥才能实施魔法。我可以帮助你，但需要你每周三都来帮我削土豆皮并制作足够多的土豆泥。这样我就能实施魔法，帮你变出一个零钱魔盒。这个零钱魔盒只能找零钱，你和奶奶不能从零钱魔盒里取钱、也不允许向零钱魔盒里放入钱，否则零钱魔盒会立即失效。注意，必须保证每周三都来，不能缺席。"

皮皮卡很爽快地答应了。后来魔法师就给皮皮卡的Java杂货店变出一个零钱魔盒。每次需要找零钱时，只要皮皮卡或奶奶对零钱魔盒说出客户买了多少钱的货品，收了顾客多少钱，零钱魔盒就立刻说出找零数额，然后盒子里就出现了需要给顾客的零钱。从此以后，零钱魔盒每次都能快速、准确地找零钱给客户，皮皮卡和奶奶的Java杂货店的生意也越来越好如图19.1所示。

图19.1 Java杂货店

本故事纯属虚构，如有雷同，纯属巧合。

19.2 场景故事的目的

1. 侧重点

零钱魔盒里一共有4个独立的小魔盒，分别是10元、5元、2元和1元的小魔盒。当需要找零钱时，如需要找零钱38元，零钱魔盒会进行如下操作：

（1）首先让10元魔盒尝试是否能完成全部或部分任务，如果只能完成部分任务，10元魔盒只能完成找零钱30元的任务（即完成了找零钱38元的部分任务）；

（2）10元魔盒把剩余的8元任务交给5元魔盒，5元魔盒完成找零钱5元的任务（即完成了找零钱8元的部分任务）；

（3）5元魔盒把剩余的3元任务交给2元魔盒，2元魔盒完成找零钱2元的任务（即完成了找零钱3元的部分任务）；

（4）2元的魔盒把剩余的1元任务交给1元魔盒，1元魔盒完成找零钱1元的任务（即完成了找零钱1元的全部任务）。

在这个过程中，零钱魔盒里的4个小魔盒共同合作完成了找零钱38元的任务。

零钱魔盒体现了面向对象中的一个重要设计理念之一：即所谓的责任链模式，其关键是将用户的请求分派给多个对象，这些对象被组织成一个责任链，即每个对象含有后继对象的引用。责任链上的一个对象，如果能完成用户的请求，就不再将用户的请求传递给责任链上的下一个对象；如果不能处理或完成用户的请求，就必须将用户的请求传递给责任链上的下一个对象。如果责任链上的末端对象也不能处理或完成用户的请求，那么用户的整个责任链就没有完成任务。

2. 涉及的其他知识点

java.util包中的Scanner类创建的对象本质上是一个输入流，可以让这个输入流指向System类的静态成员in（in是指向用户的键盘的输入流），代码如下：

```
Scanner scanner = new Scanner(System.in);
```

这样一来，scanner就可以读取用户从键盘输入的基本型数据，例如：

```
byte m = scanner.nextByte();
```

3. 进一步的尝试

给零钱魔盒增加更多的小魔盒，比如面值是5角、1角等小魔盒。

19.3 程序运行效果与视频讲解

零钱魔盒自动完成找零。主类是MainClass。程序运行效果如图19.2所示。

视频讲解

```
输入货品的钱:37
输入收取的钱:50

找零钱:10元面值1张
找零钱:5元面值0张
找零钱:2元面值1张
找零钱:1元面值1张
继续输入1,否则输入0: 1
输入货品的钱:32
输入收取的钱:100

找零钱:10元面值6张
找零钱:5元面值1张
找零钱:2元面值1张
找零钱:1元面值1张
继续输入1,否则输入0:
```

图19.2 零钱魔盒的程序运行效果

19.4 阅读源代码

（1）MoneyBox.java的代码如下：

```java
public class MoneyBox {
     public int moneyValue;                           //小魔盒的面值
     public int changeCount;                          //找零钱数
     public boolean success;                          //否找零钱成功
     private MoneyBox nextBox;                        //后继小魔盒
     public MoneyBox(int moneyValue){                 //小魔盒的面值
         this.moneyValue = moneyValue;
     }
     //用面值moneyValue的钱，把整钱money分解成小于或等于money的零钱
     public void handleChange(int money){
        int completedChangeTasks = 0;                 //本小魔盒贡献的零钱
        int n =0,sum = 0;
     //找到最多用几个面值moneyValue的钱可以把整钱分解成小于或等于money
        while(true){
           sum = sum + moneyValue;
           n++;
           if(sum > money)
             break;
        }
```

```
        n--;
        //本小魔盒可以给出的面值是moneyValue的零钱个数
        changeCount = n;
        completedChangeTasks = moneyValue*changeCount;
        if(completedChangeTasks == money) {
           success = true;        //找零钱成功
        }
        else {
           if(nextBox!= null){
             //下一个小魔盒负责处理剩余的找零钱任务
            nextBox.handleChange(money-completedChangeTasks);
           }
           else {
             success = false;     //找零钱失败
           }
        }
     }
     public void setNextMoneyBox(MoneyBox nextBox){
          this.nextBox = nextBox;
     }
}
```

（2）MagicBox.java的代码如下：

```
import java.util.Scanner;
public class MagicBox {
    MoneyBox moneyBox10;
    MoneyBox moneyBox5;
    MoneyBox moneyBox2;
    MoneyBox moneyBox1;
    MagicBox(){
       moneyBox10 = new MoneyBox(10);
       moneyBox5 = new MoneyBox(5);
       moneyBox2 = new MoneyBox(2);
       moneyBox1 = new MoneyBox(1);
       moneyBox10.setNextMoneyBox(moneyBox5);
       moneyBox5.setNextMoneyBox(moneyBox2);
       moneyBox2.setNextMoneyBox(moneyBox1);
    }
    public void giveChange(){
       Scanner scanner = new Scanner(System.in);
       System.out.print("输入货品的钱:");
       int goodsMoney =scanner.nextInt();
       System.out.print("输入收取的钱:");
```

```
int receivedMoney =scanner.nextInt();
if(goodsMoney>receivedMoney) {
   System.out.println("顾客购物钱款不足");
}
if(goodsMoney<receivedMoney) {
   moneyBox10.handleChange(receivedMoney-goodsMoney);
}
if(moneyBox10.success){
   System.out.printf(
   "\n找零钱:%d元面值%d张",moneyBox10.moneyValue,moneyBox10.
   changeCount);
}
else if(moneyBox5.success){
   System.out.printf(
   "\n找零钱:%d元面元值%d张",moneyBox10.moneyValue,moneyBox10.
   changeCount);
   System.out.printf(
   "\n找零钱:%d元面值%d张",moneyBox5.moneyValue,moneyBox5.
   changeCount);
}
else if(moneyBox2.success){
    System.out.printf(
    "\n找零钱:%d元面值%d张",moneyBox10.moneyValue,moneyBox10.
    changeCount);
    System.out.printf(
    "\n找零钱:%d元面值%d张",moneyBox5.moneyValue,moneyBox5.
    changeCount);
    System.out.printf(
    "\n找零钱:%d元面值%d张",moneyBox2.moneyValue,moneyBox2.
    changeCount);
}
else if(moneyBox1.success){
    System.out.printf(
    "\n找零钱:%d元面值%d张",moneyBox10.moneyValue,moneyBox10.
    changeCount);
    System.out.printf(
    "\n找零钱:%d元面值%d张",moneyBox5.moneyValue,moneyBox5.
    changeCount);
    System.out.printf(
    "\n找零钱:%d元面值%d张",moneyBox2.moneyValue,moneyBox2.
    changeCount);
    System.out.printf(
    "\n找零钱:%d元面值%d张",moneyBox1.moneyValue,moneyBox1.
    changeCount);
```

```
        }
        else {
            System.out.println("\n找零钱失败");
        }
    }
}
```

（3）MainClass.java的代码如下：

```
import java.util.Scanner;
public class MainClass {
   public static void main(String args[]) {
       MagicBox magicBox = new MagicBox();
       byte isContinue = 1;
       while(isContinue==1){
          magicBox.giveChange();
          Scanner scanner = new Scanner(System.in);
          System.out.print("\n继续输入1,否则输入0: ");
          isContinue = scanner.nextByte();
       }
   }
}
```

场景故事 20
苹果装箱

20.1 场景故事

Tom和Jerry管理着一个苹果园。今年，他俩要把苹果分成3个等级。

Tom对Jerry说：“我发现苹果园里的苹果的‘大小’在1~10，今年准备把‘大小’是8、9、10的苹果定为一等苹果；把‘大小’是5、6、7的苹果定为二等苹果；把‘大小’是2、3、4的苹果定为三等苹果，而‘大小’是1的苹果就不要了。”

Jerry说：“这么多苹果，咋挑拣啊？！”

Tom说：“我想到一个办法！制作一个带漏口的、具有一定长度的凹型工具，从凹型工具的一端放入苹果，让苹果滚向另一端，但是凹型工具的底部带有很光滑的漏口，漏口的大小依次是4（小于或等于4的苹果能落下）、7（小于或等于7的苹果能落下）和10（小于或等于10的苹果能落下）。这样，三等苹果和准备废弃的苹果最先从第一个漏口落到苹果箱（只要把很少的废弃苹果再挑出即可）；然后是二等苹果落到苹果箱；最后是一等苹果落到苹果箱”。

Jerry说：“真是一个好办法！”

Tom对Jerry说：“现在，需要你去制作凹槽苹果分拣机！”

凹槽苹果分拣机如图20.1所示。

图20.1　Jerry制作的凹槽苹果分拣机

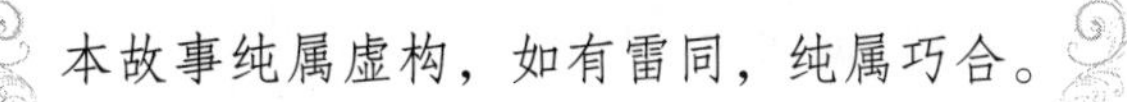

20.2 场景故事的目的

1. 侧重点

编程的一个重要理念。两个合作的对象，到底哪个负责算法？是苹果验证漏口的大小决定是否进入漏口，还是漏口验证苹果的大小决定是否让苹果漏进来？显然，是苹果要被分级，因此应当让苹果验证是否符合漏口的大小，从而决定是否进入该漏口。

在实际编程时，可方便地使用访问者模式。

漏口接收苹果：

```
public void accept(Apple apple,int boxSiz){
       apple.enterBox(this, boxSiz);
}
```

但决定是否进入漏口的算法，由苹果决定：

```
public void enterBox(AppleBox box,int boxSize) {
   ……
}
```

2. 涉及的其他知识点

集合框架中的HashSet集合。集合不允许有相同的元素，也就是说，如果b已经是集合中的元素，那么再执行set.add(b)操作是无效的（默认情况下，如果两个元素的引用相同就属于相同的元素）。

集合对象的初始容量是16字节，装载因子是0.75。换言之，如果集合添加的元素超过总容量的75%，则集合的容量将增加一倍。

使用HashSet集合遇到的特殊情况（此处苹果装箱没有遇到这种特殊情况）：两个元素的引用不同，但程序也许不希望集合中同时有这两个元素。例如，想要从电话号码中选择最后两位分别是95、85、75和65的电话号码各一串。但有许多电话号码的后两位是95，如果按照HashSet集合的默认验证条件，这些电话号码都可以添加到集合中，这就不符合只选择一个最后两位是95的电话号码的要求。这时，必须要让创建电话号码的Phone类重写从Object类继承的int hashCode()方法，让所有电话号码的“散列值”相同，即废掉 “散列值”的作用。注意，重写hashCode()方法不会影响Phone对象的引用值。这样做的原因是，当HashSet集合添加新元素时，都要首先验证集合中是否有和其“散列值”相同的元素（HashSet集合用元素的hashCode()方法返回的值作元素的“散列值”），如果没有相同的元素就直接将其添加到集合中；如果有相同的元素，再用当前元素的equals()方法（从Object类继承的方法）检查集合中是否有和当前集合相等的元素（默认情况下，如果两个元素的引用相同就属于相同的元素）。因此，Phone类可以通过重写equals()方法重新规定Phone对象相等的条件。

下列代码 WantingPhone.java演示集合中添加电话号码尾号两位分别是95、85、75和65的电话号码各一串。

WantingPhone.java的代码如下：

```
import java.util.HashSet;
import java.util.Iterator;
class Phone {
    long number;
    Phone(long n){
        number = n;
    }
   //重写hashCode方法，废掉散列值的作用：
    public int hashCode(){
        return 0;//返回任意一个int常量即可
    }
   //重写 equals方法，即重新定义对象的相等条件：
    public boolean equals(Object object){
        Phone phone =(Phone)object;
        String str1 =""+this.number;
        String str2 =""+phone.number;
        str1 = str1.substring(str1.length()-2);          //尾号后两位
        str2 = str2.substring(str2.length()-2);
        return str1.equals(str2);
    }
}
public class WantingPhone {
     public static void main(String args[]) {
         //set添加特殊尾号电话各一部
         HashSet<Phone> set=new HashSet<Phone>();
         long  number[]={ 13998727695L,17091208995L,
                          13898737685L,17171380985L,
                          13898757675L,17151618975L,
                          13698717665L,17601538965L};
         Phone phone[] = new Phone[number.length];
         for(int i=0;i<phone.length;i++){
```

```
            phone[i] = new Phone(number[i]);
            set.add(phone[i]);
        }
        System.out.println("全部电话数量:"+number.length);       //输出8
        System.out.println("特殊尾号(各一部数量):"+set.size());  //输出4
        Iterator<Phone> iter = set.iterator();
        while(iter.hasNext()){
           Phone p = iter.next();
           System.out.println(p.number);
        }
    }
}
```

3. 进一步的尝试

参照本程序，将参加百分制考试的学生分成不及格、及格、中等、良好和优秀5个等级，并给出及格率和优秀率。

20.3 程序运行效果与视频讲解

视频讲解

苹果的大小是随机形成的，所以程序每次运行效果不尽相同，主类是MainClass。程序运行效果如图20.2所示。

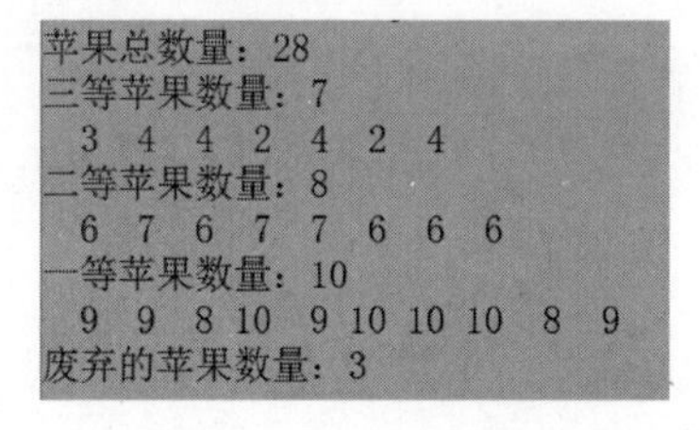

图20.2 苹果分级装箱

20.4 阅读源代码

（1）Apple.java的代码如下：

```
public class Apple {
```

```
    public int appleSize;
    public void setSize(int appleSize) {
        this.appleSize = appleSize;
    }
    public void enterBox(AppleBox box,int boxSize){
        if(this.appleSize>=boxSize&&this.appleSize-boxSize<=2)
            box.set.add(this); //向盒子添加苹果
    }
}
```

（2）AppleBox.java的代码如下：

```
import java.util.HashSet;
import java.util.Iterator;
public class AppleBox {
    public HashSet<Apple> set;
    public int boxSize;
    AppleBox(){
        set = new HashSet<Apple>();
    }
    public void setSize(int boxSiz){
        this.boxSize = boxSize;
    }
    public void accept(Apple apple,int boxSiz){
        apple.enterBox(this,boxSiz);
    }
    public int getAmount(){
        return set.size();
    }
    public Apple[] getApples(){
        Apple[] apple = new Apple[set.size()];
        Iterator<Apple> iter = set.iterator();
        int i = 0;
        while(iter.hasNext()){
            apple[i] = iter.next();
            i++;
        }
        return apple;
    }
}
```

（3）MainClass.java的代码如下：

```
import java.util.ArrayList;
```

```
import java.util.HashSet;
import java.util.Random;
public class MainClass{
   public static void main(String args[]){
      ArrayList<Apple> appleList = new ArrayList<Apple>();
      AppleBox setOne = new AppleBox();       //装一等苹果
      setOne.setSize(8);
      AppleBox setTwo = new AppleBox();
      setTwo.setSize(5);
      AppleBox setThree = new AppleBox();
      setTwo.setSize(2);
      Random random = new Random();
      int amount = 28;                       //28个苹果
      System.out.println("苹果总数量: "+amount);
      for(int i=0;i<amount;i++) {
         Apple apple = new Apple();
         apple.setSize(random.nextInt(10)+1);
                                             // "大小" 分布在1~10
         appleList.add(apple);
      }
      for(int i=0;i<appleList.size();i++){
         Apple apple = appleList.get(i);
         setThree.accept(apple,2);            //三等苹果最先被分拣出来
         setTwo.accept(apple,5);
         setOne.accept(apple,8);
      }
      System.out.println("三等苹果数量: "+setThree.getAmount());
      Apple apple[] = setThree.getApples();
      for(int i=0;i<apple.length;i++){
          System.out.printf("%3d",apple[i].appleSize);
      }
      System.out.println("\n二等苹果数量: "+setTwo.getAmount());
      apple = setTwo.getApples();
      for(int i=0;i<apple.length;i++){
          System.out.printf("%3d",apple[i].appleSize);
      }
      System.out.println("\n一等苹果数量: "+setOne.getAmount());
      apple = setOne.getApples();
      for(int i=0;i<apple.length;i++){
          System.out.printf("%3d",apple[i].appleSize);
      }
```

```
        int sum =
        setThree.getAmount()+setTwo.getAmount()+setOne.getAmount();
        System.out.println("\n废弃的苹果数量："+(amount-sum));
    }
}
```

场景故事 21

福利彩票

21.1 场景故事

“双色球”是一种由中国福利彩票发行管理中心统一组织发行的福利彩票。具体玩法：红球一共6组，每组从1~33中抽取一个，保证这6个红球互相不重复。然后，蓝球是从1~16中抽取一个数字（红球33选6乘以蓝球16选1）。一等奖（6+1）中奖概率为1/17721088 ≈0.0000056%，如图21.1所示。

图21.1　双色球福利彩票

21.2 场景故事的目的

1. 侧重点

福利彩票的过程是销售彩票、彩民购买彩票。程序需要选用某个对象存储销售的全部彩票，然后模拟产生一等奖号码，再将这个一等奖号码和存储在这个对象中的彩票进行比对，看是否有中一等奖的彩票。

这里，强一个调编程的重要理念：在某些问题中，当两个对象需要合作时，要仔细考虑应该由哪个对象负责“算法”更为合理。显然，彩民负责设置彩票上的号码信息（负责“算法”），即由彩民负责购买彩票，并确定彩票上的篮球和红球的数字。

（1）彩票允许被购买。

```
public void acceptPerson(Person p) {
      p.buyLottery(this);//由Person对象负责购买设置彩票上的号码
}
```

（2）彩民负责设置号码。

```
public void buyLottery(Lottery lottery) {
    ……
}
```

2. 涉及的其他知识点

集合框架中的HashSet集合。集合不允许有相同的元素，也就是说，如果b已经是集合中的元素，那么再执行set.add(b)操作是无效的（默认情况下，如果两个元素的引用相同就属于相同的元素）。集合对象的初始容量是16个字节，装载因子是0.75。也就是说，如果集合添加的元素超过总容量的75%时，集合的容量将增加一倍。

（1）使用HashSet集合的特殊情况。

① 两个元素的引用不同。

② 程序不希望集合中同时有这两个元素。

例如，想要从4个年级的学生中各选一名学生（即不同的元素）。但是，如果按照HashSet集合的默认验证条件，4个年级的学生都可以添加到集合中。

（2）处理办法。

① 重写hashCode()方法。

Student类重写从Object类继承的int hashCode()方法，让所有学生的散列值相同，即废掉散列值的作用（注意，重写hashCode()方法不会影响Student对象的引用值）。这样做的原因是，当HashSet集合添加新元素时，都要首先验证集合中是否有和其散列值相同的元素（HashSet集合用元素的hashCode()方法返回的值做元素的散列值）。如果没有相同的元素，就直接将其添加到集合中；如果有相同的元素，再用当前元素的equals()方法（从Object类继承的方法）检查集合中是否有和当前集合相等的元素（默认情况下，如果两个元素的引用相同，则这两元素属于相同的元素）。

② 重写equals()方法。

Student类重写equals()方法，重新规定Student对象相等的条件。集合中添加的学生分别来自4个年级别，代码如下：

```
import java.util.HashSet;
import java.util.Iterator;
class Student {
   int number;
   Student(int n){
      number = n;
   }
  //重写hashCode()方法，废掉散列值的作用
   public int hashCode(){
      return -1;                          //返回任意一个int常量即可
   }
  //重写 equals()方法，重新定义对象的相等条件：
   public boolean equals(Object object){
      Student stu =(Student)object;
      String str1 =""+this.number;
      String str2 =""+stu.number;
      str1 = str1.substring(0,4);                          //前缀4位代表年级
      str2 = str2.substring(0,4);
      return str1.equals(str2);
   }
}
public class E {
    public static void main(String args[]) {
       //set添加特殊尾号电话各一部
       HashSet<Student> set=new HashSet<Student>();
       int number[]={ 2019068,2019187,
                      2020018,2020876,
                      2021018,2021568,
                      2022762,2022987};
       Student stu[] = new Student[number.length];
       for(int i=0;i<stu.length;i++){
           stu[i] = new Student(number[i]);
           set.add(stu[i]);
       }
       System.out.println("全部学生数量:"+number.length);          //输出8
       System.out.println("4个年级各一位:"+set.size());           //输出4
       Iterator<Student> iter = set.iterator();
       while(iter.hasNext()){
```

```
            Student student = iter.next();
            System.out.println(student.number);
        }
    }
}
```

3.进一步的尝试

用文件的形式存储销售的彩票（需要使用文件输入输出流的知识）。

21.3 程序运行效果与视频讲解

其实，任何研究中彩票的方法都没什么用。因为从概率的角度来看，研究很久买一注彩票和随机地买一注彩票的中奖概率是一样的。主类是MainClass。程序运行效果如图21.2所示。实际的抽奖活动中，不会用伪随机数（计算机程序用的是伪随机数，即算法得到的随机数）。

视频讲解

```
输入销售的彩票数量（回车确认）:777777
一共销售777777注彩票
头奖红色球是：[2, 9, 12, 23, 32, 33]
头奖蓝色球是：[5]
销售777777注,没有彩票中头奖
```

(a) 销售777777注

```
输入销售的彩票数量（回车确认）:6666666
一共销售6666666注彩票
头奖红色球是：[4, 5, 8, 22, 27, 30]
头奖蓝色球是：[7]
销售6666666注,没有彩票中头奖
```

(b) 销售6666666注

图21.2　福利彩票的程序运行效果

21.4 阅读源代码

（1）Lottery.java的代码如下：

```
public class Lottery {                            //彩票
    public int [] redBallNumber = null;    //存放红球号码
    public int [] blueBallNumber = null;   //存放蓝球号码
    public void acceptPerson(Person p) {   //由Person对象负责购买这张彩票
        p.buyLottery(this);                      //由Person对象负责购买这张彩票
    }
}
```

（2）Person.java的代码如下：

```
import java.util.Random;
public class Person {                                   //买彩票的人
    public void buyLottery(Lottery lottery) {
        lottery.redBallNumber = giveRedBallNumber();
        lottery.blueBallNumber = giveBlueBallNumber();
    }
    int [] giveRedBallNumber(){
        return GiveRandomNumber.getBallNumber(6,33);
    }
    int [] giveBlueBallNumber(){
        return GiveRandomNumber.getBallNumber(1,16);
    }
}
```

（3）StartSellLottery.java的代码如下：

```
import java.util.Scanner;
import java.util.InputMismatchException;
import java.util.LinkedList;
public class  StartSellLottery {                          //销售彩票
    public static LinkedList<Lottery> sellLottery(long amountLottery) {
        LinkedList<Lottery> saveLottery = new LinkedList<Lottery>();
                                                          //存放卖出的彩票
        for(long i=0;i<amountLottery;i++) {
            Lottery lottery = new Lottery();              //创建彩票
            lottery.acceptPerson(new Person());           //彩票被人买去
            saveLottery.add(lottery);
        }
        return saveLottery;
    }
}
```

（4）GiveRandomNumber.java的代码如下：

```
import java.util.Random;
import java.util.LinkedList;
public class GiveRandomNumber {
     public static int [] getBallNumber(int count,int allCount){
        int [] ballNumber = null;        //存放随机给出的count个不同的号码
        LinkedList<Integer> saveNumber = new LinkedList<Integer>();
                                         //存放数字
        for(int i=1;i<=allCount;i++) {  //按1到allCount（球号）的顺序存
```

```
                                         //入链表saveNumber
            saveNumber.add(i);
        }
        ballNumber = new int[count];  //随机地从链表saveNumber抽取n个球
        Random random = new Random();
        int m = 0;
        while( count > 0) {
           int index = random.nextInt(saveNumber.size());
           int number = saveNumber.remove(index);  //抽取一个球不放回
           ballNumber[m] = number;  //随机得到 [1,allCount]的一个数number
           m++;
           count--;
        }
        return ballNumber;
    }
}
```

（5）MainClass.java的代码如下：

```
import java.util.LinkedList;
import java.util.Arrays;
import java.util.Iterator;
import java.util.Scanner;
public class MainClass {                     //主类，负责卖彩票和抽奖
    public static void main(String args[]) {
        long amountLottery = 0;           //销售的数量
        Scanner scanner = new Scanner(System.in);
        System.out.print("输入销售的彩票数量（回车确认）:");
        amountLottery = scanner.nextLong();
        int sum = 0;                        //头奖的彩票数量
        LinkedList<Lottery> list;         //存放卖出的全部彩票（比较大的数据）
        list = StartSellLottery.sellLottery(amountLottery);
        System.out.println("一共销售"+list.size()+"注彩票");
        int [] redBall =GiveRandomNumber.getBallNumber(6,33);
                                            //得到6个红球
        int [] blueBall =GiveRandomNumber.getBallNumber(1,16);
                                            //得到1个蓝球
        Arrays.sort(redBall);
        Arrays.sort(blueBall);
        System.out.println("头奖红色球是: "+Arrays.toString(redBall));
        System.out.println("头奖蓝色球是: "+Arrays.toString(blueBall));
        //查看list中哪个中奖了
        int n = 0;
        Iterator<Lottery> ite = list.iterator();
```

```
        while(ite.hasNext()) {
           Lottery lottery = ite.next();
           boolean isOk =
           isSame(lottery.redBallNumber,redBall)&&
           isSame(lottery.blueBallNumber,blueBall);
           if(isOk) {
                 n++;
           }
           sum = sum +n;
        }
        if(sum >=1 ){
            System.out.println(sum+"个彩票中头奖");
        }
        else {
            System.out.println("销售"+amountLottery+"注,没有彩票中头奖");
        }
    }
    static boolean isSame(int []a,int [] b){  //判断两个数组的值是否完全相同
        Arrays.sort(a);
        Arrays.sort(b);
        return Arrays.equals(a,b);
    }
}
```

场景故事 22

摆积木块

22.1 场景故事

Tom和Jerry有很多块正方体的积木，每个积木块的6个面上都各自有一个图形（例如●、★、■、◆等），如图22.1所示。有一天，Tom和Jerry决定在地上用这些积木摆出一个矩形（让矩形的长边上的积木块★朝上，短边上的积木块■朝上，4个角的积木块●朝上，矩形区域内部的积木块◆朝上）。经过一番尝试后，Jerry总觉得摆出的矩形看起来不太协调。

图22.1 Tom和Jerry的积木块

Tom说："听说有个神奇的数字——0.618，只要让矩形短边上的积木的块数是0.618分数表示的分子，长边上的积木的块数是0.618分数表示的分母，摆出的矩形就会在视觉上看起来很协调。例如，很多国家的国旗的宽和长的比是0.618。"

Jerry说："太好了！如果积木块不够用，就让矩形的长边和短边的积木块数同比例地减少就行了，Tom，你现在负责去计算0.618的分数表示！"

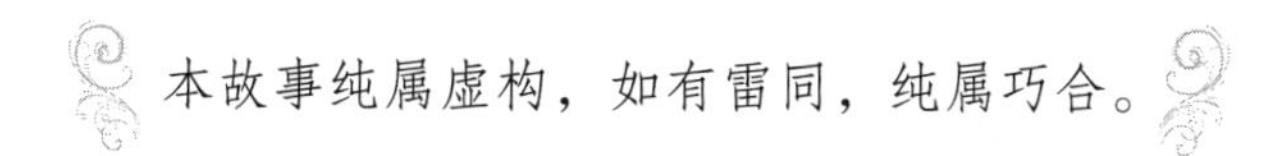

本故事纯属虚构，如有雷同，纯属巧合。

22.2 场景故事的目的

1. 侧重点

正则表达式。比如给出匹配小数的正则表达式。巩固String对象和int、double型之间的相互转换。

2. 涉及的其他知识点

String类是java.lang包中的final类，String对象中封装的字符序列是不可以再发生变化的。另外，对于：

```
String str = new String("a"+"b");
```

Java编译器在编译之前会进行优化处理，即会将"a"+"b"更改成"ab"，即将源代码更改为：

```
String str = new String("ab");
```

因此，对于String str = new String("a"+"b");字节码中会有一个常量池的常量对象"ab"和一个str变量对象，并且分配给str的字符序列存储在动态区，不在常量池。Str变量对象中的引用和"ab"常量对象的不相同。

而对于下列代码：

```
String str ="ab";
```

字节码中会有一个常量池的"ab"常量对象，一个str变量对象，但分配给str对象的字符序列存储在常量池，即str的字符序列和"ab"的相同，str中的引用和"ab"的也相同。

3. 进一步的尝试

给定一个小数，然后统计该小数部分数字出现的频率。例如，对于0.1899，数字9的频率是2/4，即50%。提示：使用String类的char charAt(int index)方法返回字符序列中的字符。

22.3 程序运行效果与视频讲解

视频讲解

将0.618表示成分数后，用分子、分母同比例缩小后的数目作为矩形的长边和短边上的积木块数。主类是MainClass。程序运行效果如图22.2所示。

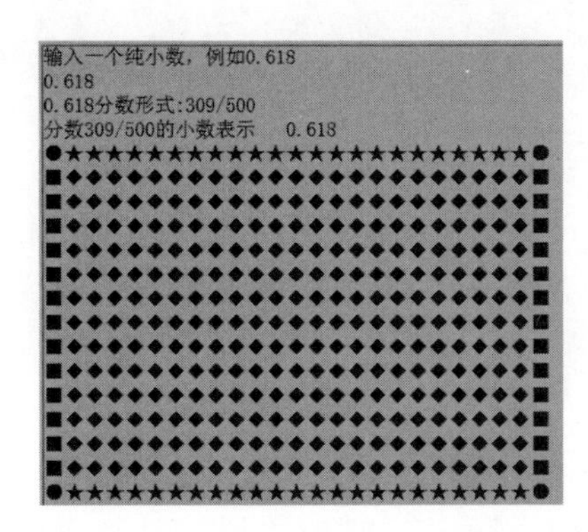

图22.2　根据0.618的分子、分母摆设的矩形

22.4　阅读源代码

（1）Fraction.java的代码如下：

```
public class Fraction {
    public int fenzi = 1;
    public int fenmu = 1;
     //求小数decimal的分数表示
    public void setFraction(double decimal){
        String numberString = String.valueOf(decimal);
        String xiaoshuPart =
        numberString.substring(numberString.indexOf(".")+1);
                                                      //得到小数部分
        int m = xiaoshuPart.length();//m的值就是小数的小数位数
        fenzi = Integer.parseInt(xiaoshuPart); //分子
        fenmu = (int)Math.pow(10,m);           //分母
        int b = f(fenzi,fenmu);                //分子、分母的最大公约数
        fenzi = fenzi/b;
        fenmu = fenmu/b;
    }
    static int f(int a,int b) {                //求a和b的最大公约数
      if(a==0) return 1;
      if(a<b) {
         int c = a;
         a = b;
         b = c;
```

```
        }
        int r=a%b;
        while(r!=0) {
           a = b;
           b = r;
           r = a%b;
        }
        return b;
    }
}
```

（2）MainClass.java的代码如下：

```
import java.util.Scanner;
public class MainClass {
    public static void main(String args[]) {
        String regex = "0.[0-9]+";  //匹配任何小数形式的正则表达式
        Scanner scanner = new Scanner(System.in);
        System.out.println("输入一个纯小数,例如0.618");
        double d = scanner.nextDouble();
        String numberString = String.valueOf(d);
        while(!numberString.matches(regex)) {
            System.out.println(numberString+"不是纯小数,重新输入");
            scanner = new Scanner(System.in);
            d = scanner.nextDouble();
            numberString = String.valueOf(d);
        }
        Fraction fraction = new Fraction();
        fraction.setFraction(d);                    //计算小数d的分数表示
        int fenzi = fraction.fenzi;
        int fenmu = fraction.fenmu;
        System.out.println(d+"分数形式:"+fenzi+"/"+fenmu);
        d = fenzi/(double)fenmu;
        System.out.printf("分数%d/%d的小数表示%8.3f",fenzi,fenmu,d);
        int m = fenmu/20;
        int n = fenzi/20;
        System.out.println();
        System.out.print("●");                      //矩形的角
        for(int i=2;i<=m-1;i++){
           System.out.print("★");                   //矩形的长边
        }
        System.out.print("●");                      //矩形的角
        System.out.println();
        for(int i =2;i<=n-1;i++){
```

```
            System.out.print("■");                    //矩形的短边
            for(int j =2;j<=m-1;j++){
              System.out.print("◆");                  //矩形的内部
            }
            System.out.print("■");
            System.out.println();
        }
        System.out.print("●");                        //矩形的角
        for(int i =2;i<=m-1;i++){
            System.out.print("★");
        }
        System.out.print("●");                        //矩形的角
    }
}
```

场景故事 23

神秘的蛋糕

23.1 场景故事

Tom和Jerry正在屋子里追逐嬉闹，突然听到敲门声，Tom开门一看，原来是送信的邮差。

Tom和Jerry打开信一看，上面写道："你好，Tom和Jerry。我是一个魔法师，我将于明天上午给你们送去一块蛋糕和一份朗读材料，朗读材料的内容都是用#号分隔的单个字母，而且这些字母的顺序也毫无规律。如果你们谁能更快地把朗读材料中的字母读给我听，并说出字母的个数，这块蛋糕就归谁！注意，信的背面有秘籍。"

正当Tom想看看信的背面时，被手疾眼快的Jerry，一把抢走并迅速跑开了。

Tom心想："我追了你Jerry这么多年，还怕你跑？！"于是，Tom直奔Jerry追去，就在Jerry即将进入自己的小屋洞口之前，Tom一把抓住了Jerry。

Jerry不慌不忙地一挥手说："且慢！让我看一遍信背面的内容。"Jerry查看信的内容如图23.1所示。

图23.1　Jerry查看信的内容

然后Tom就夺回了信，看到信的背面写着："秘籍：此信在手。"

第二天一大早，魔法师真的来了，手里拿着一个大蛋糕。魔法师进屋后说：

“这个游戏规则是我一共最多会抛出8次硬币，如果是正面朝上，则Tom先开始读字母给我听，然后是Jerry；如果是反面朝上，则Jerry先开始读字母给我听，然后是Tom。在8次比赛中，谁获胜5次谁就是获胜者，但手中持有我寄来的书信的人，在8次比赛中只须获胜两次就可以获胜；如果两人都是获胜者，则二人平分蛋糕。”

Tom紧紧握住手中的信，高兴地说道；“快开始吧！我已经迫不及待地想吃蛋糕了！”

Jerry一声不响，非常淡定地摇晃着手里用来吃蛋糕的叉子和盘子。

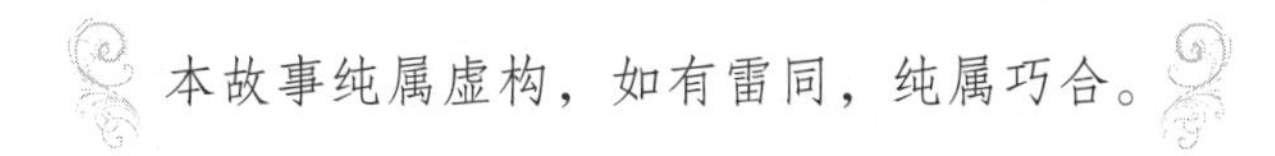

23.2 场景故事的目的

1. 侧重点

StringTokenizer类和Scanner类都可用于分解字符序列中的单词，但二者在思想上有所不同。

StringTokenizer类把分解出的全部单词都存放到StringTokenizer对象的实体中。因此，StringTokenizer对象能较快地获得单词（StringTokenizer对象用空间换取时间）。

Scanner类不会把单词存放到Scanner对象的实体中，而是仅存放怎样获取单词的分隔标记。因此，Scanner对象获得单词的速度相对较慢，但使用Scanner对象的好处是可以节省内存空间（即用时间换取空间）。如果字符序列存放在磁盘空间的文件中，并且形成的文件比较大，那么用Scanner对象分解字符序列中的单词就可以节省内存。

StringTokenizer对象一旦产生就会立刻知道单词的数目，即可以使用countTokens()方法返回单词的数目。而Scanner类不能提供这样的方法，因为Scanner类不会把单词存放到Scanner对象的实体中，如果想知道单词的数目，就必须去一个一个地获取并记录单词的数目。

2. 涉及的其他知识点

FileInputStream类创建的对象被称为文件字节输入流。输入流调用read()方法顺序地读取源中的单个字节数据，该方法返回字节值（0~255的一个整数）。如果到达源的末尾，该方法返回-1。

3. 进一步的尝试

让Tom和 Jerry读取英文单词。

23.3 程序运行效果与视频讲解

视频讲解

魔法师寄来的信为一个文本文件letter.txt，如图23.2所示。

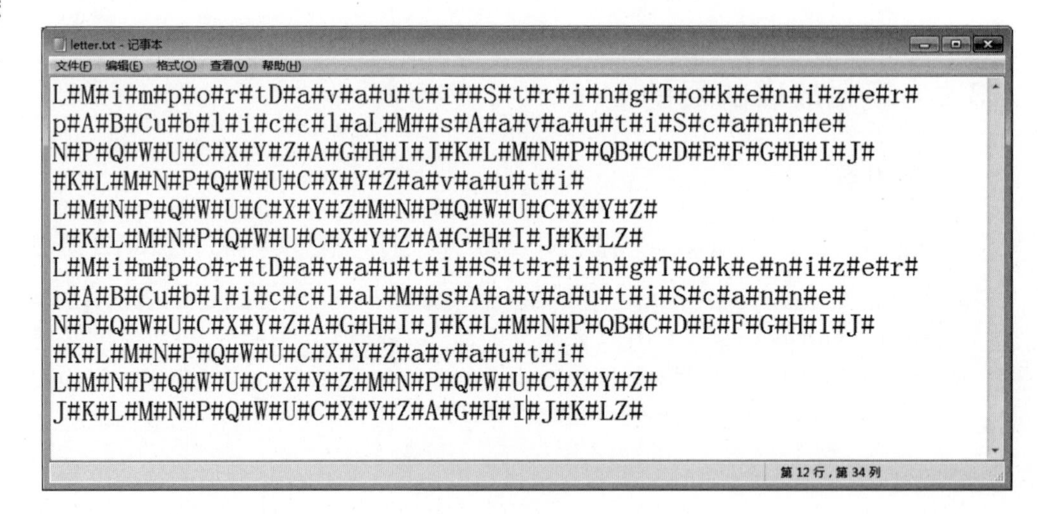

L#M#i#m#p#o#r#tD#a#v#a#u#t#i##S#t#r#i#n#g#T#o#k#e#n#i#z#e#r#
p#A#B#Cu#b#l#i#c#c#l#aL#M##s#A#a#v#a#u#t#i#S#c#a#n#n#e#
N#P#Q#W#U#C#X#Y#Z#A#G#H#I#J#K#L#M#N#P#QB#C#D#E#F#G#H#I#J#
#K#L#M#N#P#Q#W#U#C#X#Y#Z#a#v#a#u#t#i#
L#M#N#P#Q#W#U#C#X#Y#Z#M#N#P#Q#W#U#C#X#Y#Z#
J#K#L#M#N#P#Q#W#U#C#X#Y#Z#A#G#H#I#J#K#LZ#
L#M#i#m#p#o#r#tD#a#v#a#u#t#i##S#t#r#i#n#g#T#o#k#e#n#i#z#e#r#
p#A#B#Cu#b#l#i#c#c#l#aL#M##s#A#a#v#a#u#t#i#S#c#a#n#n#e#
N#P#Q#W#U#C#X#Y#Z#A#G#H#I#J#K#L#M#N#P#QB#C#D#E#F#G#H#I#J#
#K#L#M#N#P#Q#W#U#C#X#Y#Z#a#v#a#u#t#i#
L#M#N#P#Q#W#U#C#X#Y#Z#M#N#P#Q#W#U#C#X#Y#Z#
J#K#L#M#N#P#Q#W#U#C#X#Y#Z#A#G#H#I#J#K#LZ#

图23.2 魔法师寄来的信

多次运行程序发现，Jerry获胜次数多于Tom。主类是MainClass。程序运行效果如图23.3所示。

(a) Tom和Jerry同时获胜　　(b) Jerry获胜

图23.3 Tom与Jerry的比赛结果

23.4 阅读源代码

（1）Jerry.java的代码如下：

```
import java.util.StringTokenizer;
import java.io.FileInputStream;
import java.io.File;
import java.io.IOException;
public class Jerry {
    StringTokenizer fenxi;
    File jerryFile = null;
    String str = null;                              //存放文件的内容
    public  void read(File file) {
       if(jerryFile != file) {
           jerryFile = file;
           str = readFile();
       }
       fenxi = new StringTokenizer(str,"#"); //单词封装在此fenxiww对象里
       int count = fenxi.countTokens();
       while(fenxi.hasMoreTokens()) {
           String item = fenxi.nextToken();
           System.out.printf("%3s",item);
       }
       System.out.printf("\n字母个数:"+count);
    }
    private String readFile(){
       StringBuffer buffer=new StringBuffer();
       try{
         FileInputStream in = new FileInputStream(jerryFile);
          int m =-1;
          while((m=in.read())!=-1){
            char c = (char)m;
            buffer.append(c);
          }
       }
       catch(IOException exp){}
       return new String(buffer);
  }
}
```

（2）Tom.java的代码如下：

```
import java.util.Scanner;
```

```
import java.io.File;
import java.io.FileNotFoundException;
public class Tom {
    Scanner scanner;
    public void read(File file) {
       try{
         scanner = new Scanner(file);          //单词没封装在scanner对象里
         scanner.useDelimiter("[#]+");         //scanner设置分隔标记
       }
       catch(FileNotFoundException exp){}
       int count = 0;
       while(scanner.hasNext()){
          String item = scanner.next();
          count++;
          System.out.printf("%3s",item);
       }
       System.out.printf("\n字母个数:"+count);
    }
}
```

（3）MainClass.java的代码如下：

```
import java.util.Random;
import java.util.ArrayList;
import java.util.Collections;
import java.io.File;
public class MainClass {
    public static void main(String args[]) {
        File letter = new File("letter.txt");
        Tom tom = new Tom();
        Jerry jerry = new Jerry();
        Random random = new Random();
        boolean ok = false;                         //模拟抛硬币
        long timingTom = 0;                         //存放计时
        long timingJerry = 0;
        int winTom = 0;                             //Tom赢的次数
        int winJerry = 0;
        int countLoop = 1;
        while(countLoop<=8) {
           ok = random.nextBoolean();
           if(ok) {
              System.out.println("\nTom开始读");
              long startTime = System.nanoTime();
              tom.read(letter);
```

```
                long endTime = System.nanoTime();
                timingTom = endTime-startTime;
                System.out.printf("\nJerry开始读");
                startTime = System.nanoTime();
                jerry.read(letter);
                endTime = System.nanoTime();
                timingJerry = endTime-startTime;
                if(timingTom<timingJerry){
                    ++winTom;
                }
                else {
                    ++winJerry;
                }
            }
            else {
                System.out.println("\nJerry开始读");
                long startTime = System.nanoTime();
                jerry.read(letter);
                long endTime = System.nanoTime();
                timingJerry = endTime-startTime;
                System.out.println("\nTom开始读");
                startTime = System.nanoTime();
                tom.read(letter);
                endTime = System.nanoTime();
                timingTom = endTime-startTime;
                if(timingTom<timingJerry){
                    ++winTom;
                }
                else {
                    ++winJerry;
                }
            }
            countLoop++;
        }
        if(winJerry>=5&&winTom<2){
            System.out.printf("\n%s","Jerry获胜");
        }
        if(winJerry>=5&&winTom>=2){       //Tom只要两次获胜即可吃到蛋糕
          System.out.printf("\n%s","Jerry和Tom同时获胜");
        }
        if(winTom>=2&&winJerry<5){
            System.out.printf("\n%s","Tom获胜",winTom);
        }
        System.out.printf("\n%s:Tom%5d,Jerry%5d","获胜次数",winTom,winJerry);
    }
}
```

场景故事 24

女友的生日

24.1 场景故事

Tom和女友相处有一段时间了，Tom曾问过女友的生日是哪一天，可他的女友就是不说。有一天，女友突然对Tom说：“我要过生日了，我的几个好朋友也想见见你，生日那天你也来吧，地点就是我们俩认识后第一次一起吃饭的地方，时间是晚上18点。”

Tom高兴地问女友：“你的生日是哪一天呢？”

女友给了他一个纸条，条上写着两串数字6673673和33333333。然后女友说：“一定准时来哦，否则…”

Tom瞠目结舌地看着6673673和33333333，不知所措。女友给的纸条，如图24.1所示。Tom突然想起了好友Jerry，他聪明又是计算机专业的，决定去咨询Jerry。

图24.1　女友给Tom的纸条

Jerry看了看纸条上的6673673和33333333，然后说：“这事好办！你放心，我马上就知道你女友的生日，但我现在不能告诉你，等到了那一天我自然会告诉你，不会耽误你的幸福大事。但是需要你从今天起，每天来我家一次，帮我打扫房间一个小时。”

Tom想起了女友可爱的脸庞，只好默默地点头答应，并按照Jerry说的去做。过了几天，Jerry果然打电话给Tom，Tom按时参加了女友的生日Party，大家都很高兴。女友也一直夸奖Tom聪明。

本故事纯属虚构，如有雷同，纯属巧合。

24.2 场景故事的目的

1. 侧重点

String类、StringBuffer类、正在表达式。

2. 涉及的其他知识点

java.util包中的ArrayList<E>泛型类。ArrayList<E>泛型类声明和创建数组表时，必须要指定泛型E的具体类型，例如：

```
ArrayList<Long> saveRemainder;
saveRemainder = new ArrayList<Long>();
```

3. 进一步的尝试

使用正则表达式判断循环小数的循环部分是不是一个日期。

24.3 程序运行效果与视频讲解

程序运行后，输入6673673/33333333（一个无限循环小数），可以看出女友的生日是2002年10月19日。主类是MainClass。程序运行效果如图24.2所示。

视频讲解

```
按b/a格式输入一个纯正分数，例如6/7
6673673/33333333
分数6673673/33333333的小数信息是:
0.20021019 20021019 20021019...
从第1位出现循环数字序列
循环的数字序列是
20021019
循环的数字序列长度
8
```

图24.2　女友的生日的程序运行效果

24.4　阅读源代码

（1）Fraction.java的代码如下：

```
import java.util.ArrayList;
public class Fraction {
     long b = -1;                         //存放分子
     long a = -1;                         //存放分母
     StringBuffer saveDecimal;            //保存小数中的数字序列
     ArrayList<Long> saveRemainder;  //保存计算过程中的余数
     public String loopDogit;             //保存无限循环小数中那些循环的数字序列
     boolean islimited;                   //是否是有限小数
     boolean isLoop;                      //是否是无限循环小数
     long position = -1;                  //存放小数开始出现无限循环的位置
     long decimalDigit = -1;              //临时存放计算出的小数位上的数字
     public void setDenominator(long a) {        //设置分母
         this.a = a;
     }
     public void setNumerator(long b) { //设置分子
         this.b = b;
     }
     public Fraction(){
        saveDecimal = new StringBuffer();
        saveRemainder = new ArrayList<Long>();
     }
     public void handle() {
         islimited = isLoop = false;
         saveDecimal = saveDecimal.delete(0,saveDecimal.length());
         saveRemainder.clear();        //清空存放的内容
         saveRemainder.add(b);
         long r = -1;
         while(true){
```

```
            long n = findN();           //看看分子b乘以几次10才能大于分母a
            decimalDigit = b/a;         //注意，b已经发生变化
            if(n == 1){
               saveDecimal.append(decimalDigit);
               //把计算出的小数位上的数字尾加到saveDecimal
            }
            else if(n>1){
               for(long i = 0;i<n-1;i++){
                 saveDecimal.append("0");  //小数部分需要额外添加n-1个0
                saveRemainder.add(-1L);     //-1不参与计算的余数
               }
               saveDecimal.append(decimalDigit);
            }
            r = b%a;
            if(r == 0) {
                  islimited = true;          //是有限小数
                  break;
            }
            int index = -1;
            if(saveRemainder.contains(r)) {
                  isLoop = true;             //是无限循环小数
                  index = saveRemainder.indexOf(r);
                  position = index;     //保存开始循环的小数数字的位置
                  loopDogit = saveDecimal.substring(index);
                  break;
            }
            saveRemainder.add(r);        //把计算过程中的余数尾加到saveushu
            b = r;
         }
   }
   public String getDecimalMess(){
      String str = null;
      if(isLoop){
         str =
      new String("0."+saveDecimal+" "+loopDogit+" "+loopDogit+"...");
      }
      if(islimited){
         str =  new String("0."+saveDecimal);
      }
      return str;
   }
   long findN() {
      long n = 1;
      while(true) {
```

```
            b = b*10;
            if(b>=a) {
                break;
            }
            n++;
        }
        return n;
    }
}
```

（2）MainClass.java的代码如下：

```
import java.util.Scanner;
public class MainClass {
    public static void main(String args[]) {
        String regex = "[1-9][0-9]*/[1-9][0-9]*";     //匹配任何正分数
        Scanner scanner = new Scanner(System.in);
        System.out.println("按b/a格式输入一个纯正分数,例如6/7");//13/23
        long b = 1;                                       //存放分子
        long a = 1;                                       //存放分母
        Fraction fraction = new Fraction();
                                    //等待用户从键盘输入形如a/b格式的序列
        String fractionStr = scanner.nextLine();
        String [] str = fractionStr.split("/");    //分解出b/a格式中的b和a
        b = Integer.parseInt(str[0]);
        a = Integer.parseInt(str[1]);
        long greatestDivisor = f(a,b);             //分母、分子的最大公约数
        b = b/greatestDivisor;
        a = a/greatestDivisor;
        boolean isFraction = false;
        if(b<a)
            isFraction = true;                     //是真分数
        while(isFraction == false) {
            System.out.println(fractionStr+"不是真分数,重新输入");
            scanner = new Scanner(System.in);
            fractionStr = scanner.nextLine();
            str = fractionStr.split("/");
            b = Integer.parseInt(str[0]);
            a = Integer.parseInt(str[1]);
            greatestDivisor = f(a,b);              //分母、分子的最大公约数
            b = b/greatestDivisor;
            a = a/greatestDivisor;
            if(b<a)
                isFraction = true;                 //是真分数
```

```
        }
        fraction.setNumerator(b);
        fraction.setDenominator(a);
        fraction.handle();                                  //处理分数
        System.out.println("分数"+b+"/"+a+"的小数信息是：");
        System.out.println(fraction.getDecimalMess());
        if(fraction.isLoop){
            System.out.println("从第"+(fraction.position+1)+"位出现循环数
            字序列");
            System.out.println("循环的数字序列是\n"+fraction.loopDogit);
            System.out.println("循环的数字序列长度\n"+fraction.loopDogit.
            length());
        }
    }
    static long f(long a,long b) {                          //求a和b的最大公约数
      if(a==0) return 1L;
      if(a<b) {
         long c = a;
         a = b;
         b = c;
      }
      long r=a%b;
      while(r!=0) {
         a = b;
         b = r;
         r = a%b;
      }
      return b;
    }
}
```

神奇的数字1

25.1 场景故事

很久以前，有兄弟两人，老大叫Tom，老二叫Jerry。一天，村里来了个教书先生，兄弟二人决定跟着他读书学习。很快4年过去了，教书先生说："Tom和Jerry做事都很细致。今天，老师就出一道题目，考考你们兄弟二人。我曾经给你们讲过黄金分隔数0.618是一个无理数。自然界里很多优美的图案里都蕴含着这个数字，比如人的身体，如果上身的长度比下身的长度接近0.618，那么这个人的身材看起来就会非常好。但人们都不知道这个无理数的准确值，也许只能通过自然界里某些优美的图形来感受它的存在。例如，人们通过圆感觉圆周率。"老师接着说："我只给你们一个数字1，你俩从1出发，计算黄金分隔数，看用谁的办法计算出的黄金分割数的精度更高。"

（1）Tom使用的算法。

Tom在《经典三十六个算法》学过一个算法：

令item等于1，然后按下列步骤进行计算。

第1步，将item赋值给result :result = item。

第2步，将1/(1+item)赋值给item: item=1/(1+item)。

第3步，再回到第1步。

如此反复多次，result的值就会接近黄金分隔数。Tom决定用此算法来计算。

（2）Jerry使用的算法。

Jerry在《二十四个算法秘籍》学过一个算法：

令F1，F2等于1，然后按下列步骤进行计算。

第1步，将F1/F2赋值给result：result = F1/F2。

第2步，将F1+F2 赋值给F1：F1 = F1+F2，接着将F2+F1赋值给F2。

第3步，回到第1步。

如此反复多次，result的值就会接近黄金分隔数。Jerry决定用此算法来计算，如图25.1所示。

图25.1 Tom和Jerry的算法

25.2 场景故事的目的

1. 侧重点

MathContext类和BigDecimal类。当需要更精准地处理带小数点的数时（如需要精度为大于16位的有效数字），可以使用BigDecimal类。基本用法如下:

（1）定义precision精度。使用java.math包中的MathContext类的对象定义精度。例如,创建的precision对象定义的精度是100位，采用四舍五入：

```
MathContext precision = new MathContext(100,RoundingMode.HALF_UP);
```

RoundingMode类中的许多static常量定义了精度采用的取舍办法（比如也有五舍六入）。

（2）按照precision精度创建BigDecimal类的对象，例如，四则运算：

```
BigDecimal oneNumber = new BigDecimal("3.567678",precision);
BigDecimal twoNumber = new BigDecimal("5.980899",precision);
BigDecimal result =null;
result = oneNumber.add(twoNumber,precision);
result = oneNumber.subtract(twoNumber,precision);
result = oneNumber.multiply(twoNumber,precision);
result = oneNumber.divide(twoNumber,precision);
```

2. 涉及的其他知识点

无限分式。形如（a是正整数）：

a/(a+(a/(a+(a/(a+/⋯)

称为无限分式，无限分式的值是无理数。当a等于1时，无限分式是黄金分隔数。

3．进一步的尝试

计算无限分式的近似值，其中a的值可以从键盘输入。

25.3　程序运行效果与视频讲解

视频讲解

程序运行后，Tom的算法和Jerry的算法在保留100位精度的情况下，仅仅在第100位有误差，可见二位的算法精度极其接近。主类是MainClass。程序运行效果如图25.2所示。

```
Tom计算黄金分隔数:
精度是precision=100 roundingMode=HALF_UP位的黄金分隔数
0.6180339887498948482045868343656381177203091798057628621354486227052604628189024497072072041893911377
Jerry计算黄金分隔数:
精度是precision=100 roundingMode=HALF_UP位的黄金分隔数
0.6180339887498948482045868343656381177203091798057628621354486227052604628189024497072072041893911375
```

图25.2　算法的精度

25.4　阅读源代码

（1）Tom168.java的代码如下：

```
import java.math.MathContext;
import java.math.RoundingMode;
import java.math.BigDecimal;
public class Tom168 {
    static public String get168ByBigDecimal(int n) {
        //规定计算精度保留n位有效数字（四舍五入）
        MathContext precision =
        new MathContext(n,RoundingMode.HALF_UP);
        BigDecimal result =null;
        BigDecimal one = new BigDecimal("1",precision);
        BigDecimal item = one;
        int i=1;
        while(i<=99999){
```

```
            result = item;
            item =
            one.divide(one.add(result,precision),precision);
            i++;
        }
        return "精度是"+precision+"位的黄金分隔数\n"+result.toString();
    }
}
```

（2）Jerry168.java的代码如下：

```
import java.math.MathContext;
import java.math.RoundingMode;
import java.math.BigDecimal;
public class Jerry168 {
    static public String get168ByBigDecimal(int n) {
        //规定计算精度保留n位有效数字（四舍五入）
        MathContext precision =
        new MathContext(n,RoundingMode.HALF_UP);
        BigDecimal result =null;
        BigDecimal one = new BigDecimal("1",precision);
        BigDecimal F1= one;
        BigDecimal F2= one;
        BigDecimal item = F1.divide(F2,precision);
        int i=1;
        while(i<=99999){
            result = F1.divide(F2,precision);;
            F1 = F1.add(F2,precision);
            F2 = F2.add(F1,precision);
            i++;
        }
        return "精度是"+precision+"位的黄金分隔数\n"+result.toString();
    }
}
```

（3）MainClass.java的代码如下：

```
public class MainClass {
    public static void main(String args[]){
        System.out.println("Tom计算黄金分隔数:");
        System.out.println(Tom168.get168ByBigDecimal(100));
        System.out.println("Jerry计算黄金分隔数:");
        System.out.println(Jerry168.get168ByBigDecimal(100));
    }
}
```

会Java能脱单

26.1 场景故事

Tom是个性格很开朗的人，但一直也都没有谈过恋爱。有一天，Tom的老朋友Jerry来到Tom家，对Tom说：“我给你介绍个对象吧，女孩是Java村的人，长得美丽大方、端庄秀气。”

Tom一听是Java村的女孩，心里很高兴，因为方圆百里的乡亲们都说Java村的女孩人美心又好。

第二天，Jerry就安排Tom和女孩见了面。

事后，Jerry问Tom是否喜欢她。Tom说：“很好啊，可就是不知道人家喜不喜欢我？”

Jerry对Tom说：“人家女孩不像你，她有点害羞，但她说会给你写信的，你就回家等女孩的信吧！”

过了几天，Tom竟然真的收到了女孩的来信，如图26.1所示。Tom打开信一看，心里凉了半截，尽管Tom的英文不太好，但是《新概念英语》第二册第1篇课文还是很熟悉的，姑娘的来信的内容就是一篇英文课文，而且姑娘还把课文中某些脍炙人口的句子做了特殊处理，把I said angrily.故意写成：

```
I said angrilyyyyyyyyyyyyyyyyyyyyyyyyyyyyyyyyyyyyyyyyyyyyyyyyyyyyyy.
```

把It’s none of your business.故意写成：

```
It’s none of your businessssssssssssssssssssssssssssssssss.
```

把This is a private conversation!故意写成：

```
This is a private conveeeeeeeeeeeeeeeeeeeeeeeeeersation!
```

TOM看到这些，心里彻底凉透了。

Tom理解的是：可能前几天的约会，他的表现让女孩很生气、很不满意，认为女孩很不喜欢他，用书信表达了自己拒绝的态度。

正当Tom躺在床上闷闷不乐时，听到了有人敲门的声音。Tom开门一看，是Jerry来了。

Jerry看了信，悄悄地对Tom说了几句话后，Tom破涕为笑、心情一下子好了起来，然后对Jerry说："谢谢我的老朋友，我马上写程序，然后给女孩回信！"

图26.1　姑娘给Tom的神秘来信

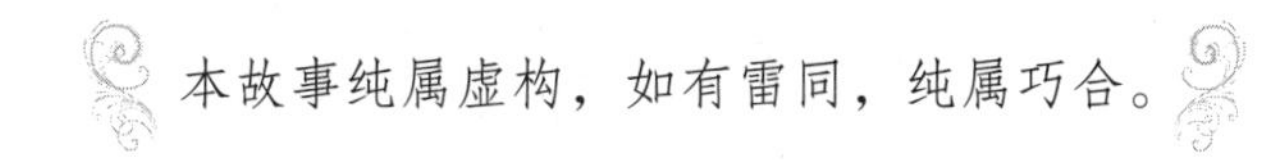

26.2　场景故事的目的

1．侧重点

Jerry偷偷对Tom说的话是：你需要统计出英文字母的出现频率，按频率从高到低的顺序，查看前3个英文字符是不是yes或前两个英文字符是不是no。如果前3个英文字符是yes，你就给她回信，把频率表发给女孩，并附留言：Nice to meet you。

本场景故事目的是巩固用输入流读取文件的知识，重点体会怎样使用对象封装基本需求中的数据，然后用面向对象的方法使用算法变得简单。

2．涉及的其他知识点

ArrayList<E>实现了泛型接口List<E>，ArrayList<E>泛型类的对象采用顺序结构来存储数据，习惯上称ArrayList类创建的对象为数组表。使用ArrayList<E>泛型类声明创数组表时，必须要指定E的具体类型，然后数组表就可以使用add(E obj)方法向数组表依次增加节点。

3. 进一步的尝试

统计英文文章中单词的频率。提示：可以使用java.util包中的Scanner类读取文件。

```
Scanner scanner = new Scanner(new File("English.txt"));
String regex = "[\\s\\d\\p{Punct}]+";//匹配空格，标点符号等组成的字符序列
Scanner.useDelimiter(regex);
```

26.3 程序运行效果与视频讲解

视频讲解

Tom运行自己写的程序后，发现女孩的回答果然是yes。主类是MainClass。程序运行效果如图26.2所示。

```
y频率10.49   e频率 9.90   s频率 7.20   n频率 6.45   t频率 6.15
a频率 6.00   i频率 5.40   o频率 4.80   r频率 3.60   d频率 3.30
h频率 2.25   u频率 1.95   l频率 1.80   g频率 1.50   w频率 1.20
v频率 1.05   c频率 0.90   m频率 0.90   p频率 0.60   k频率 0.45
b频率 0.45   j频率 0.15   f频率 0.15
    ________________
  y      e      s
    ________________
```

图26.2 会Java能脱单的程序运行效果

26.4 阅读源代码

（1）CF.java的代码如下：

```
public class CF {
    char ch;                              //存放一个英文字符
    int count;                            //存放ch出现的次数
    double frequency;                     //存放频率
    public void setChar(char c) {
        ch = c;
```

```
    }
    public char getChar() {
        return ch;
    }
    public void setCount(int n) {
        count = n;
    }
    public int getCount(){
        return count;
    }
    public void setFrequency(double d){
        frequency = d;
    }
    public double getFrequency(){
        return frequency;
    }
    //重写Object类的方法，规定对象相等
    public boolean equals(Object o) {
        CF other = (CF)o;
        return this.ch == other.ch;
    }
}
```

（2）Jerry168.java的代码如下：

```
import java.util.ArrayList;
import java.io.FileInputStream;
import java.io.File;
import java.io.IOException;
import java.util.Collections;
public class Frequency {
    ArrayList<CF> list = null;
    double contentLength;
    Frequency(){
       list = new ArrayList<CF>();
    }
    public void setContent(File file){
        list.clear();
        try{
          FileInputStream in = new FileInputStream(file);
          contentLength = file.length();
          int m =-1;
          while((m=in.read())!=-1){
            char c = (char)m;
```

```
            if(Character.isUpperCase(c))
               c = (char)(c+32);                    //变小写
            CF cf = new CF();
            if(c>='a'&&c <='z'){
               cf.setChar(c);
               if(!list.contains(cf)){
                  list.add(cf);
                  cf.setCount(1);
               }
               else {
                 CF tempCF= list.get(list.indexOf(cf));
                 tempCF.setCount(tempCF.getCount()+1);
               }
            }
         }
      }
      catch(IOException exp){}
      computerFrequency();
   }
   public void computerFrequency(){
      for(int i=0;i<list.size();i++){
         CF cf = list.get(i);
         double m = cf.getCount();
         double frequency = (m/contentLength)*100;
         cf.setFrequency(frequency);
      }
   }
   public void showFrequency(){
      Collections.sort(list,(CF a,CF b)->{return b.count-a.count;});
      for(int i=0;i<list.size();i++){
         CF cf = list.get(i);
         System.out.printf
        ("%5c频率%5.2f",cf.getChar(),cf.getFrequency());
         if((i+1)%5==0)
           System.out.printf("\n");
      }
   }
   public void showTopThree(){
     System.out.printf("%s","\n\t----------------\n");
     for(int i=0;i<3;i++){
        CF cf = list.get(i);
        System.out.printf("%7c",cf.getChar());
     }
     System.out.printf("%s","\n\t----------------");
```

```
    }
}
```

（3）MainClass.java的代码如下：

```
import java.io.File;
public class MainClass {
    public static void main(String args[]) {
        File file = new File("hello.txt");
        Frequency frequency = new Frequency();
        frequency.setContent(file);
        frequency.showFrequency();
        frequency.showTopThree();                //看看是不是yes
    }
}
```

场景故事 27

报恩的蚂蚁

27.1 场景故事

很久以前，有一个叫Jerry的农民，因家里很穷，只好给Tom地主家干活。某天天空还下着雨，Jerry在田间边上发现了一个蚂蚁窝，旁边的蚂蚁们正在忙碌地搬家。可是，河水马上就要淹没蚂蚁的窝，Jerry急忙用土筐将蚂蚁窝搬到了安全的地方。

转眼到了年底，按照事先的约定，Tom要给Jerry 20斗大米和15斗小米作为报酬。当Jerry来到Tom家时，发现Tom家的大米和小米都混在一起，堆成了一个大堆。

Jerry迷惑不解地问Tom：“你为何要这样做？”

Tom说：“我的妻子喜欢吃‘二米饭’，她让我直接把大米、小米混在一起，免得做饭时总是忘记放小米或忘记放大米！”

Jerry为难地对Tom说：“我的那份大米和小米也混堆里了吗？”

Tom说：“是啊，你自己慢慢挑拣吧，挑好了喊我就行了。”

这可愁坏了Jerry，但Jerry也只好慢慢地一粒一粒的挑拣。过了几个时辰，Jerry的身边来了数不清的蚂蚁，开始帮着他挑拣大米和小米。过了不一会儿，蚂蚁就将大米和小米分成了两堆， 然后蚂蚁就消失了。蚂蚁帮助Jerry分拣大米和小米，如图27.1所示。

图27.1　蚂蚁帮助Jerry分拣大米和小米

Jerry把Tom喊过来， Tom看后大吃一惊，并将Jerry的20斗大米和15斗小米给Jerry。Jerry走了以后，Tom又默默地将大米和小米混成了一堆，没办法，他的妻子就爱吃“二米饭”。

27.2 场景故事的目的

1. 侧重点

（1）巩固利用对象检索所需要的数据。例如，从一个文本文件中检索出整数，浮点数和日期（类似蚂蚁帮助Jerry挑选大米）。

（2）巩固Pattern类和Match类有关的知识点（二者在java.util.regex包中）。

（3）一个重要的设计思想。当某个对象想完成任务时，可以把任务分派给其他对象来帮助完成。例如，将查找数据的方法封装在接口中，实现接口的类去完成查找工作（策略模式是体现这一重要设计思想的模式之一，称实现接口的类为一种策略）。

① Pattern对象。

使用正则表达式得到Pattern对象pattern：

```
Pattern pattern = Pattern.compile(regex);
```

② Matcher对象。

String对象input得到Matcher对象：

```
Matcher matcher = pattern.matcher(input);
```

③ Matcher对象的常用方法。

Matcher对象matcher调用各种方法检索input：

例如，Matcher对象matcher依次调用booelan find()方法检索input的字符序列中和regex匹配的子字符序列。

2. 涉及的其他知识点

java.lang包中的StringBuffer对象封装字符序列和操作字符序列的许多方法。StringBuffer对象封装的字符序列是可被修改的，StringBuffer类能提供了可修改字符序列的许多方法。

3. 进一步的尝试

编写检索E-mail地址或手机电话号码的类。

27.3 程序运行效果与视频讲解

视频讲解

从文本文件（data.txt）：

武汉长江大桥于1955-9-1动工，1957年10月15日正式通车。大桥总长1670m，其中，主桥长1156m，跨度分别为17.2m和16.0m，下层铁路桥宽14.5m，桥上两侧人行道宽2.25m，总投资额为1.38亿元人民币。武汉的水位比国家基准面低1.66m（通常所说的海拔高-1.66m）。武汉市平均海拔为23.3m。武汉最冷的记录-10℃。文稿日期：2020/03/31。

其中，检索整数（不包括浮点数和日期中的数字）、浮点数和日期。主类是MainClass。程序运行效果如图27.2所示。

```
一共有3个整数
[1670, 1156, -10]

一共有8个浮点数
[17.2, 16.0, 14.5, 2.25, 1.38, 1.66, -1.66, 23.3]

一共有3个日期
[1955-9-1, 1957年10月1日, 2020/03/31]
```

图27.2　检索整数、浮点数和日期

27.4 阅读源代码

（1）AntFindData.java的代码如下：

```
import java.io.File;
public interface AntFindData {
    public String[] findData(File file);
}
```

（2）Person.java的代码如下：

```
import java.io.File;
public class Person {
    AntFindData ant;
```

```
    public void setAntFindData(AntFindData ant){
       this.ant = ant;
    }
    public String [] findingData(File file){
       String result[] = ant.findData(file);
       return result;
    }
}
```

（3）FindFloat.java的代码如下：

```
import java.util.regex.Pattern;
import java.util.regex.Matcher;
import java.util.Arrays;
import java.util.ArrayList;
import java.io.File;
import java.io.FileInputStream;
import java.io.IOException;
public class FindFloat implements AntFindData {
      Pattern pattern;                                   //模式对象
      Matcher matcher;                                   //匹配对象
      String dataSource;                                 //要检索的数据源
      FindFloat(){
          String regex="-?[0-9][0-9]*[.]+[0-9]*" ;     //匹配浮点数，不包括整数
          pattern = Pattern.compile(regex);            //初始化模式对象
      }
      public String[] findData(File file){
         StringBuffer buffer = new StringBuffer();
         byte [] b= new byte[100];
         try{
            FileInputStream in = new FileInputStream(file);
            int m =-1;
            while((m=in.read(b))!=-1){
                String str = new String(b,0,m);
                buffer.append(str);
            }
            dataSource = new String(buffer);
          }
          catch(IOException exp){
             System.out.println("无数据源");
             return null;
          }
          return find();
      }
```

```
    private String[] find(){     //初始化匹配对象，用于检索dataSource
        matcher = pattern.matcher(dataSource);
        ArrayList<String> list = new ArrayList<String>();
        while(matcher.find()) {
           String str = matcher.group();
           list.add(str);
        }
        String [] str =new String[list.size()];
        for(int i=0;i<str.length;i++){
           str[i] = list.get(i);
        }
        return str;
    }
}
```

（4）FindDate.java的代码如下：

```
import java.util.regex.Pattern;
import java.util.regex.Matcher;
import java.util.Arrays;
import java.util.ArrayList;
import java.io.File;
import java.io.FileInputStream;
import java.io.IOException;
public class FindDate implements AntFindData {
    Pattern pattern;                                        //模式对象
    Matcher matcher;                                        //匹配对象
    String dataSource;                                      //要检索的数据源
    FindDate(){
        String year = "[1-9][0-9]{3}";
        String month = "((0?[1-9])|(1[012]))";
        String day = "((0?[1-9])|([12][0-9])|(3[01]?))";
        String regex = year+"[-年/]"+month+"[-月/]"+day;
        pattern = Pattern.compile(regex);                   //初始化模式对象
    }
    public String[] findData(File file){
       StringBuffer buffer = new StringBuffer();
       byte [] b= new byte[100];
       try{
          FileInputStream in = new FileInputStream(file);
          int m =-1;
          while((m=in.read(b))!=-1){
              String str = new String(b,0,m);
              buffer.append(str);
```

```
            }
            dataSource = new String(buffer);
        }
        catch(IOException exp){
            System.out.println("无数据源");
            return null;
        }
        return find();
    }
    private String[] find(){
        //初始化匹配对象，用于检索dataSource
        matcher = pattern.matcher(dataSource);
        ArrayList<String> list = new ArrayList<String>();
        while(matcher.find()) {
            String str = matcher.group();
            list.add(str);
        }
        String [] str =new String[list.size()];
        for(int i=0;i<str.length;i++){
            str[i] = list.get(i);
        }
        return str;
    }
}
```

（5）FindInteger.java的代码如下：

```
import java.util.regex.Pattern;
import java.util.regex.Matcher;
import java.util.Arrays;
import java.util.ArrayList;
import java.io.File;
import java.io.FileInputStream;
import java.io.IOException;
public class FindInteger implements AntFindData {
    Pattern pattern;                              //模式对象
    Matcher matcher;                              //匹配对象
    String dataSource;                            //要检索的数据源
    FindInteger(){
        String regex ="-?[0-9][0-9]*" ;           //匹配整数
        pattern = Pattern.compile(regex);         //初始化模式对象
    }
    public String[] findData(File file){
       StringBuffer buffer = new StringBuffer();
```

```
        byte [] b= new byte[100];
        try{
          FileInputStream in = new FileInputStream(file);
          int m =-1;
          while((m=in.read(b))!=-1){
              String str = new String(b,0,m);
              buffer.append(str);
          }
          dataSource = new String(buffer);
        }
        catch(IOException exp){
           System.out.println("无数据源");
           return null;
        }
        return find();
    }
    private String[] find(){
        //初始化匹配对象，用于检索dataSource
        matcher = pattern.matcher(dataSource);
        ArrayList<String> list = new ArrayList<String>();
        while(matcher.find()) {
           String str = matcher.group();
          int indexEnd = matcher.end();
                                   //返回检索到的字符序列的结束位置后面的位置
           int indexStart = matcher.start();
                                   //返回检索到的字符序列的开始位置
           boolean isFloat =
           dataSource.charAt(indexEnd)=='.'||
           dataSource.charAt(indexStart>=1?indexStart-1:0)=='.';
           boolean isDate =
           dataSource.charAt(indexEnd)=='年'||
           dataSource.charAt(indexEnd)=='月'||
           dataSource.charAt(indexEnd)=='日'||
           dataSource.charAt(indexEnd)=='-' ||
           dataSource.charAt(indexEnd)=='/' ||
           dataSource.charAt(indexStart-1)=='/' ||
           dataSource.charAt(indexStart)=='-'&&
           dataSource.charAt(indexEnd)=='-' ||
           dataSource.charAt(indexStart)=='-'&&
           Character.isDigit(dataSource.charAt(indexStart-1));
           if(!isDate&&!isFloat) {
               list.add(str);                    //不要日期和浮点数里的数字
           }
        }
```

```
        String [] str =new String[list.size()];
        for(int i=0;i<str.length;i++){
           str[i] = list.get(i);
        }
        return str;
    }
}
```

（6）MainClass.java的代码如下：

```
import java.io.File;
import java.util.Arrays;
public class MainClass {
    public static void main(String args[ ]){
        File file = new File("data.txt");
        Person jerry = new Person();
        jerry.setAntFindData(new FindInteger());
        String result[] =jerry.findingData(file);
        System.out.printf("\n一共有%d个整数\n",result.length);
        System.out.println(Arrays.toString(result));
        jerry.setAntFindData(new FindFloat());
        result =jerry.findingData(file);
        System.out.printf("\n一共有%d个浮点数\n",result.length);
        System.out.println(Arrays.toString(result));
        jerry.setAntFindData(new FindDate());
        result =jerry.findingData(file);
        System.out.printf("\n一共有%d个日期\n",result.length);
        System.out.println(Arrays.toString(result));
    }
}
```

场景故事 28

恋爱时光

28.1　场景故事

一转眼，Tom和Java村的女孩谈恋爱也有一段日子了。这对恋人如胶似漆，感情很好，两人也都有结婚的打算。一天，Tom准备好礼物，打算向女孩求婚。令他意想不到的是，女孩向他提出了几个问题。

女孩问："Tom，还记得我们开始谈恋爱的日子吗？"

Tom回答说："当然记得啊！"

接着，女孩要求Tom说出从谈恋爱那天开始，到当前求婚的日期和时间，一共经历了多少天、多少个月、多少个星期、多少小时、多少分、多少秒？

Tom顿时目瞪口呆，回答不上来，如图28.1所示。

图28.1　Tom求婚遇到的时间差考题

女孩见Tom为难的样子，便说："好吧，那你明天上午11:58再来求婚，并回答刚才的问题吧！"女孩又告诉Tom关于计算的一些计算的技巧：

（1）计算年时，不足一年的月份要省略，比如去年的8月到今年10月，那么就算一年。

（2）计算月时，要按照实际天数计算，方法是看起始月是哪一月，并且要看这个月一共有多少天，比如起始月是2月， 2月一共有28天；然后再看下一个月，下一个月是3月，那么就再过31天算一个月，最后不足一个月的天数就不在要求之内了。

（3）星期、小时、分计算方法同理。

女孩对Tom说："明天11:58见，别让我失望哦！"

Tom回到家里，拿出笔和纸立刻开始计算，可是他忙了半天，也没有算出来。Tom只好又去求助好朋友Jerry。

Jerry对Tom说："听说Java村的java.time酒吧的老板叫LocalDateTime，非常擅长计算与时间有关的问题，你可以去找他帮忙，说不定很快就能帮你搞定！"

Tom对Jerry感激不尽，Tom说："Jerry，你真是见多识广，我这就去请教。"

第2天，Tom按时求婚，并获得成功！

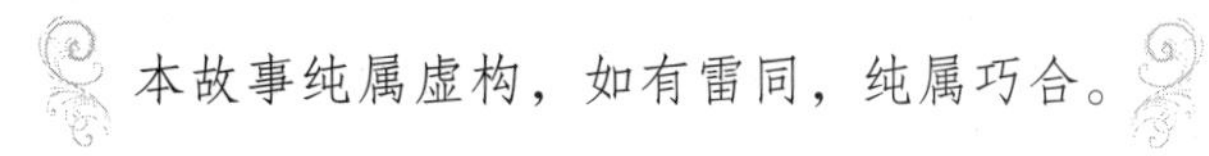

28.2 场景故事的目的

1. 侧重点

在JDK 8 之后的版本中，可以使用java.time包中的一些类，处理与时间有关的问题，比如计算两个日期差。

（1）LocalDateTime类。

可以将与日期有关的数据封装在 LocalDateTime类的实例（对象）中，例如2018-9-18,10:36:56：

```
LocalDateTime dateStart = LocalDateTime.of(2018,9,18,10,36,56);
LocalDateTime dateEnd = LocalDateTime.of(2020,5,30,18,12,29);
```

封装当前计算机时钟的时间：

```
LocalDateTime currentTime = LocalDateTime.now();
```

（2）计算日期差。

LocalDateTime对象可以使用LocalDataTime类提供的下列方法：

```
public long until(java.time.temporal.Temporal time,
```

```
java.time.temporal.TemporalUnit unit)
```

方法返回当前日期（LocalDateTime对象）和参数日期time的差值，差值依赖于参数unit，即依赖单位。unit可以取值ChronoUnit枚举类中的枚举常量，例如：

```
ChronoUnit.SECONDS
ChronoUnit.MINUTES
ChronoUnit.HOURS;
ChronoUnit.DAYS;
ChronoUnit.WEEKS;
ChronoUnit.MONTHS;
ChronoUnit.YEARS;
```

例如，对于：

```
LocalDateTime dateStart = LocalDateTime.of(2018,9,18,10,36,56);
LocalDateTime dateEnd = LocalDateTime.of(2020,5,30,18,12,29);
```

（1）相差的年。

```
long years = dateStart.until(dateEnd,ChronoUnit.YEARS);
```

那么，years的值就是1（years的值不是2，until 在计算dataStart和dateEnd差时，不足1年的零头按0处理）。

（2）相差的月。

```
long months= dateStart.until(dateEnd,ChronoUnit MONTHS);
```

months的值是20（不足1月的零头按0处理）。

2. 涉及的其他知识点

格式化日期。String类的format方法：

```
format(格式化模式,日期列表);
```

按照“格式化模式”返回“日期列表”中所列各个日期中所含数据。可以在“格式化模式”中使用“<”，比如"%tY-%<tm-%<td"中的3个格式符将格式化为同一日期：

```
String str = String.format("%tY年%<tm月%<td日",time);
```

格式符%tY、%tm和%td将分别表示日期中的“年”“月”和“日”。%ty将“年”格式化为两位数，%tY将“年”格式化为4位数。

3. 进一步的尝试

输出Tom求婚时的日期，并输出该日期是当年的第几天？该年是否是闰年？

28.3 程序运行效果与视频讲解

给出Tom开始恋爱的时间和Tom求婚的时间。主类是MainClass。程序运行效果如图28.2所示。

视频讲解

```
Tom开始恋爱的时间:
2018-09-18,10:05:08
Tom求婚的时间:
2021-05-30,11:58:00
按年算2年有余
按月算32个月有余
按天算985天有余
按星期算140个星期有余
按小时算23641小时有余
按分钟算1418512分钟有余
按秒算85110772秒

Tom从恋爱到求婚经历了:
2年零8个月零12天，零1小时52分52秒
Tom求爱成功!
```

图28.2　Tom的恋爱时光的程序运行效果

28.4 阅读源代码

MainClass.java的代码如下：

```java
import java.time.LocalDateTime;
import java.time.temporal.ChronoUnit;
public class MainClass{
   public static void main(String args[ ]) {
        LocalDateTime dateStart =
        LocalDateTime.of(2018,9,18,10,05,08);
        System.out.println("Tom开始恋爱的时间：");
        String dataFormat =
        String.format("%tY-%<tm-%<td,%<tH:%<tM:%<tS",dateStart);
        System.out.println(String.format(dataFormat));
        LocalDateTime requestMarriage =
```

```
LocalDateTime.of(2021,05,30,11,58,00);
System.out.println("Tom求婚的时间: ");
dataFormat =
String.format("%tY-%<tm-%<td,%<tH:%<tM:%<tS",requestMarriage);
System.out.println(String.format(dataFormat));
long years =
dateStart.until(requestMarriage,ChronoUnit.YEARS);
System.out.printf("按年算%d年有余\n",years);
long months=
dateStart.until(requestMarriage,ChronoUnit.MONTHS);
System.out.printf("按月算%d个月有余\n",months);
long days=
dateStart.until(requestMarriage,ChronoUnit.DAYS);
System.out.printf("按天算%d天有余\n",days);
long weeks=
dateStart.until(requestMarriage,ChronoUnit.WEEKS);
System.out.printf("按星期算%d个星期有余\n",weeks);
long hours=
dateStart.until(requestMarriage,ChronoUnit.HOURS);
System.out.printf("按小时算%d小时有余\n", hours);
long minutes=
dateStart.until(requestMarriage,ChronoUnit.MINUTES);
System.out.printf("按分钟算%d分钟有余\n",minutes);
long seconds=
dateStart.until(requestMarriage,ChronoUnit.SECONDS);
System.out.printf("按秒算%d秒\n",seconds);
//计算零头月
dateStart = dateStart.plusYears(years);
long remainderMonth =
dateStart.until(requestMarriage,ChronoUnit.MONTHS);
//计算零头天
dateStart = dateStart.plusMonths(remainderMonth);
long remainderDay =
dateStart.until(requestMarriage,ChronoUnit.DAYS);
//计算零头小时
dateStart = dateStart.plusDays(remainderDay);
long remainderHour =
dateStart.until(requestMarriage,ChronoUnit.HOURS);
//计算零头分钟
dateStart = dateStart.plusHours(remainderHour);
long remainderMinute =
dateStart.until(requestMarriage,ChronoUnit.MINUTES);
//计算零头秒
dateStart = dateStart.plusMinutes(remainderMinute);
```

```
        long remainderSecond =
        dateStart.until(requestMarriage,ChronoUnit.SECONDS);
        System.out.println("\nTom从恋爱到求婚经历了:");
        System.out.printf
        ("%d年零%d个月零%d天，零%d小时%d分%d秒\n",
        years,remainderMonth,remainderDay,remainderHour,remainderMinute,
        remainderSecond);
        System.out.println("Tom求婚成功！");
    }
}
```

场景故事 29

三盒苹果的风波

29.1 场景故事

Jerry、Tom和Spike是好朋友。某天，3人坐在Tom家的沙发上看电视，突然听到有人敲门的声音，开门一看，是魔法师。这次，魔法师带来了3盒苹果，每盒的品牌各不相同，分别叫ArrayList、LinkedList和TreeSet。魔法师给Tom、Jerry和Spike看了这3个盒子。他们发现这3种盒子里各装满了100个苹果，但每个盒子的结构却有着很大的不同，如图29.1所示。

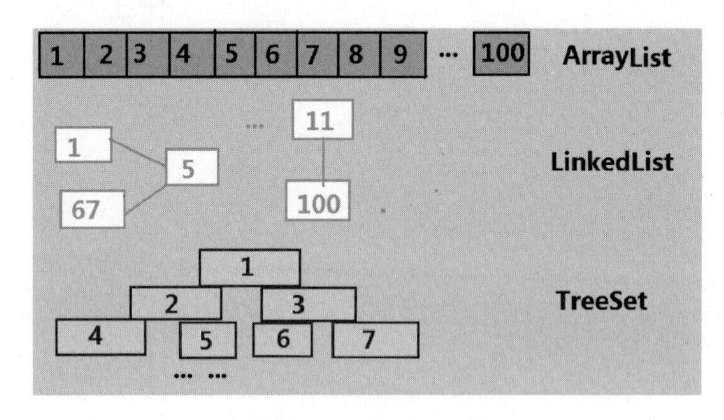

图29.1　结构不同的盒子

（1）ArrayList盒子里的格子是按号码的顺序排成一排。

（2）LinkedList盒子里的格子是乱序摆放的，但也可以慢慢地找到格子，因为1号格子用一根小小的细绳连接5号格子，5号格子用细绳连接着67号。最后是11号格子连接着100号格子。

（3）TreeSet盒子里的格子是呈树状摆放的，Tom、Jerry和Spike还注意到苹果都是由大到小的顺序放入格子里的，1号格子里的苹果最小，100号格子里的苹果最大。

魔法师说："Tom、Jerry和Spike可以各自选一盒苹果，然后立刻把最大的苹果拿出来，如果谁速度最慢，我就收回他的盒子，当然也就吃不到苹果了。"

Tom手疾眼快，拿了LinkedList盒子，因为Tom发现LinkedList盒子里的苹果更大些；Spike拿了TreeSet盒子；Jerry最后别无他选，只能拿ArrayList盒子，这个盒子里的苹果看起来是3个盒子里最小的了。

那么，大家猜猜看，最后到底是谁没吃到苹果呢？

29.2 场景故事的目的

1. 侧重点

巩固集合框架的知识点。程序可能经常需要对用于实现List结构中的数据按照某种大小关系排序。Collections类提供的用于排序的方法public static sort(List<E> list)，可以将List中的对象升序排列。List实例中存放的对象可以通过实现Comparable接口，即通过实现该接口中的方法int compareTo(Object b)来规定对象的大小关系。TreeSet实例能自动地对树节点中的对象按从小到大的顺序排列（一层一层地依次排列）。TreeSet实例中存放的对象可以通过实现Comparable接口（java.lang包中的接口）来规定对象的大小关系。

2. 涉及的其他知识点

System类的public static long nanoTime()方法只能用于测量经过时间（以纳秒为单位），并且与系统或挂钟时间的任何其他概念无关。返回的值表示自某个任意的、固定的起始时间以来的纳秒数。例如，要测量某些代码所需的执行时间：

```
long startTime = System.nanoTime();
 //...正在测量的代码...
long elapsedNanos = System.nanoTime() - startTime;
```

3. 进一步的尝试

再多一个品牌叫Stack苹果。Stack<E>实现了泛型接口List<E>，Stack<E>泛型类的对象采用栈式结构存储数据（先进后出），习惯上称Stack类创建的对象为堆栈。使用Stack<E>泛型类声明创建堆栈时，必须要指定E的具体类型，然后堆栈就可以使用void push(E item)，从而实现压栈操作，使用public E pop()方法实现弹栈操作。

29.3 程序运行效果与视频讲解

视频讲解

Jerry的ArrayList盒子里的格子井然有序，只要知道1号格子在哪里，就能立刻算出其他号的盒子在哪里。

Tom的LinkedList盒子里的格子是乱序摆放的，Tom必须从1号格子开始，即由1号格子找到5号格子，然后再由5号格子找到后续的格子，直到找到100号格子。

Spike的TreeSet盒子里的格子是呈树状排列的，last()方法使用了散列函数，用关键字查找最后一个格子（使用NavigableMap接口）。所以，通常来说，Tom没有Jerry的速度快，除非Tom运气好，比如，刚好1号格子的绳子连接的下一个格子就是100号格子。Spike的last()方法和Tom的getLast()方法的效率差不多，很难说分出伯仲。主类是MainClass。多次运行程序，发现Tom和Spike经常吃不到苹果。主类是MainClass。程序运行效果如图29.2所示。

```
Jerry开始从盒子里取苹果:
Jerry用时6158毫秒。
Tom开始从盒子里取苹果:
Tom用时116177毫秒。
Spike开始从盒子里取苹果:
Spike用时96883毫秒。
Tom没吃到苹果
```

(a) Tom没吃到苹果

```
Jerry开始从盒子里取苹果:
Jerry用时8621毫秒。
Tom开始从盒子里取苹果:
Tom用时19705毫秒。
Spike开始从盒子里取苹果:
Spike用时21347毫秒。
Spike没吃到苹果
```

(b) Spike没吃到苹果

图29.2　看谁没吃到苹果

29.4 阅读源代码

（1）Apple.java的代码如下：

```
public class Apple implements Comparable<Apple>{
   public int size;
   public Apple(int size){
      this.size = size;
   }
   public int compareTo(Apple a){
      Apple apple =a;
      return size - apple.size;
   }
}
```

（2）MainClass.java的代码如下：

```
import java.util.ArrayList;
import java.util.LinkedList;
import java.util.TreeSet;
import java.util.Collections;
import java.util.Iterator;
public class MainClass {
    public static void main(String args[]){
        long max =0;
        int appleAmount = 1000;
        LinkedList<Apple>appleTom = new LinkedList<Apple>();
        for(int i=1;i<= appleAmount;i++){
            appleTom.add(new Apple(i));
        }
        Collections.sort(appleTom);
        TreeSet<Apple>appleSpike = new TreeSet<Apple>();

        for(int i=1;i<= appleAmount;i++){
            appleSpike.add(new Apple(i));
        }
        ArrayList<Apple>appleJerry = new ArrayList<Apple>();
        for(int i=1;i<= appleAmount;i++){
            appleJerry.add(new Apple(i));
        }
        Collections.sort(appleJerry);
        System.out.println("Jerry开始从盒子里取苹果:");
        long startTime = System.nanoTime();
        appleJerry.get(appleJerry.size()-1);
        long elapsedNanosJerry = System.nanoTime()-startTime;
        if(elapsedNanosJerry>max)
           max = elapsedNanosJerry;
        System.out.printf("Jerry用时%d毫秒。\n",elapsedNanosJerry);
        System.out.println("Tom开始从盒子里取苹果:");
        startTime = System.nanoTime();
        appleTom.getLast();
        long elapsedNanosTom = System.nanoTime()-startTime;
        if(elapsedNanosTom>max)
           max = elapsedNanosTom;
        System.out.printf("Tom用时%d毫秒。\n",elapsedNanosTom);
        System.out.println("Spike开始从盒子里取苹果:");
        startTime = System.nanoTime();
        appleSpike.last();
        long elapsedNanosSpike = System.nanoTime()-startTime;
```

```
        System.out.printf("Spike用时%d毫秒。\n",elapsedNanosSpike);
        if(elapsedNanosSpike>max)
           max = elapsedNanosSpike;
        if(max == elapsedNanosTom)
           System.out.println("Tom没吃到苹果");
        if(max ==  elapsedNanosJerry)
           System.out.println("Jerry没吃到苹果");
        if(max == elapsedNanosSpike)
           System.out.println("Spike没吃到苹果");
    }
}
```

钓鱼比赛

30.1 场景故事

Tom躺在沙发上听广播，听到一件令他非常高兴的消息：附近的Python村要举办一次钓鱼活动，鱼不按大小论价，只按种类、条数计算价格。例如，草鱼的价格是26元/条，鲫鱼的价格是19元/条等。Tom心想，那要是钓着大鱼可就赚了啊。广播里还播出了一条特别令Tom心动的活动规则：在规定的时间内，钓上来的鱼的总价如果达到某个数值，就可以免费带走自己钓的鱼，否则就要买走自己所钓的鱼。可是，Python村对参赛者有个额外的要求，即参赛者必须以团队形式参加，每队人数不少于3人，按队计算所钓的鱼的总价格。Tom立刻约Jerry和Spike一起去Python村参加钓鱼活动，如图30.1所示。可是Jerry和Spike都表示对钓鱼并不是很感兴趣。

图30.1　Tom约Jerry和Spike去垂钓

Tom只好说，如果能得到免费的鱼，就请Spike吃烤肉，请Jerry吃蛋糕。

Jerry冷冷地对Tom说：“如果得不到免费鱼呢？”

Tom说：“我出钱，买下你俩钓的全部鱼！”

本故事纯属虚构，如有雷同，纯属巧合。

30.2 场景故事的目的

1. 侧重点

举类型（鱼的种类）、线程（垂钓者）以及集合框架中的某些类（用于模拟存放鱼）。如果需要一种类型的变量的取值范围包含有限个值，而且这些值名（字符序列）很少再发生变化，就可以考虑用枚举类型。使关键字enum定义枚举类型。定义枚举类型时也可以声明变量和定义方法。声明了一个枚举类型后，就可以用该枚举类型的枚举名声明一个枚举变量，例如：

```
Season time;
```

声明了一个枚举变量time。枚举变量time只能取值枚举类型中的常量，通过使用枚举名和“.”运算符获得枚举类型中的常量，例如：

```
time = Season.spring;
```

示例如下：

```
enum Season {
   spring,summer,autumn,winter;//4个枚举常量
   public int getDays(){
      int days = 0;
      if(this == spring) {
          days = 92;
      }
      else if(this == summer) {
         days = 92;
      }
      else if(this == autumn) {
         days = 91;
      }
      else if(this == winter) {
         days = 91;
      }
      return days;
   }
}
public class MainClass{
   public static void main(String args[]){
      Season season;
      season = Season.spring;
```

```
        System.out.println(season.getDays());
        System.out.printf("%s",season);
    }
}
```

2. 涉及的其他知识点

编程的一个重要理念：两个合作的对象，到底哪个负责算法？即咬钩的算法是由鱼塘里的鱼负责，还是由垂钓者负责。显然，咬钩的权利应该属于鱼塘里的鱼。在实际编程时，可方便地使用访问者模式：

（1）Persom类负责访问鱼塘。

```
void fishing(){
  pond.acceptFishing(this);//咬钩权利交给鱼塘里的鱼
}
```

（2）Fishpond类负责算法。

```
public void void acceptFishing(Person person){
      ……//具体算法
}
```

3. 进一步的尝试

给鱼（FishNameAndPrice）增加double型的属性weight，使得程序能输出垂钓者所钓之鱼的总重量。

30.3 程序运行效果与视频讲解

垂钓时，鱼是否咬钩由鱼自己决定，而且其算法是随机算法，因此每次程序运行的效果也不尽相同。主类是MainClass。程序运行效果如图30.2 所示。

视频讲解

```
汤姆鱼桶里的鱼：
 鲫鱼:19元, 草鱼:26元, 草鱼:26元。
斯派克鱼桶里的鱼：
 鲤鱼:12元, 鲤鱼:12元, 鲫鱼:19元, 鲤鱼:12元, 鲫鱼:19元, 草鱼:26元。
杰瑞鱼桶里的鱼：
 鲫鱼:19元, 草鱼:26元, 鲤鱼:12元。
汤姆成绩:71元
杰瑞成绩:57元
斯派克成绩:100元
Tom队的成绩:228元
Tom请spike吃烤肉，请jerry吃蛋糕
鱼鱼塘里剩下的鱼数量：116条
```

图30.2 Tom队取得的成绩

30.4 阅读源代码

（1）FishNameAndPrice.java的代码如下：

```
public enum FishNameAndPrice {                              //定义枚举类型
    鲤鱼,鲫鱼,草鱼;                                          //3个枚举常量
    int price;                                              //鱼的价格
    public void setPrice(int p){
        price = p;
    }
    public int getPrice(){
        return price;
    }
}
```

（2）Fishpond.java的代码如下：

```
import java.util.LinkedList;
import java.util.Random;
public class Fishpond {                                     //鱼塘
    LinkedList<FishNameAndPrice> fishInPond;    //存放鱼塘里的鱼(枚举类型)
    Random random;
    Fishpond(){
       fishInPond =new LinkedList<FishNameAndPrice>();
       random = new Random();
    }
    public void setFishAmount(int amount){                  //设置鱼的数量
        FishNameAndPrice fish = null;                       //fish是枚举类型
        for(int i=0;i<amount;i++) {
            if(i%3==0) {
                fish = FishNameAndPrice.鲤鱼;
                fish.setPrice(12);
                fishInPond.add(fish);
            }
            else if(i%3==1){
               fish = FishNameAndPrice.鲫鱼;
               fish.setPrice(19);
               fishInPond.add(fish);
            }
            else if(i%3==2){
               fish = FishNameAndPrice.草鱼;
               fish.setPrice(26);
               fishInPond.add(fish);
            }
```

```
        }
    }
    public void acceptFishing(Person person) {
        FishNameAndPrice fish = null;
        int n = random.nextInt(15);
        if(n == 3||n==7||n==12){
           if(!fishInPond.isEmpty()){
             fish = fishInPond.removeFirst();
             person.fishBucket.add(fish);
           }
        }
    }
    public int getFishAmount(){
        return fishInPond.size();
    }
}
```

（3）Person.java的代码如下：

```
import java.util.ArrayList;
public class Person extends Thread {
     Fishpond pond;
     ArrayList<FishNameAndPrice> fishBucket;  //存放垂钓的鱼
     Person(){
        fishBucket = new ArrayList<FishNameAndPrice>();
     }
     public void setFishpond(Fishpond pond){
        this.pond = pond;
     }
     void fishing(){
       pond.acceptFishing(this);                //咬钩权利交给鱼塘pond里的鱼
     }
     public int getFishingPrice(){
        int priceSum = 0;
        for(int i=0;i<fishBucket.size();i++){
           FishNameAndPrice fish =fishBucket.get(i);
           int price = fish.getPrice();
           priceSum += price;
        }
        return priceSum;
     }
     public void showFishing(){
        System.out.println(""+getName()+"鱼桶里的鱼：");
        for(int i=0;i<fishBucket.size();i++){
```

```
            FishNameAndPrice fish =fishBucket.get(i);
            System.out.printf("%3s:%2d元,",fish,fish.getPrice());
        }
        System.out.println();
    }
    public void run(){
        for(int i = 0;i<30;i++){                //30次垂钓机会
            System.out.printf("\n"+getName()+"在钓鱼中");
            fishing();
            try {
                Thread.sleep(200);
            }
            catch(InterruptedException exp){}
        }
    }
}
```

（4）MainClass.java的代码如下：

```
import java.util.ArrayList;
public class MainClass{
    public static void main(String args[]){
        int MAX = 100;                          //垂钓100元以上，免费
        Fishpond competitionPond;               //比赛鱼塘
        competitionPond = new Fishpond();
        competitionPond.setFishAmount(128);     //鱼塘里有128条鱼
        Person tom,jerry,spike;                 //垂钓者
        tom = new Person();
        tom.setName("汤姆");
        jerry = new Person();
        jerry.setName("杰瑞");
        spike = new Person();
        spike.setName("斯派克");
        tom.setFishpond(competitionPond);
        spike.setFishpond(competitionPond);
        jerry.setFishpond(competitionPond);
        tom.start();                            //开始垂钓
        jerry.start();
        spike.start();
        while(true) {                           //主线程检查垂钓者是否钓鱼完毕
            if(!tom.isAlive()&&!jerry.isAlive()&&!spike.isAlive())
                break;
        }
        System.out.println("---------");
```

```
        tom.showFishing();
        spike.showFishing();
        jerry.showFishing();
        int sum = tom.getFishingPrice()+
                  jerry.getFishingPrice()+
                  spike.getFishingPrice();
        System.out.println(""+tom.getName()+"成绩:"+tom.
        getFishingPrice()+"元");
        System.out.println(jerry.getName()+"成绩:"+jerry.
        getFishingPrice()+"元");
        System.out.println(spike.getName()+"成绩:"+spike.
        getFishingPrice()+"元");
        System.out.println("Tom队的成绩:"+sum+"元");
        if(sum>=MAX) {
          System.out.println("Tom请Spike吃烤肉,请Jerry吃蛋糕");
        }
        else {
          System.out.println("Tom付费买鱼");
        }
        System.out.println
        ("鱼塘里剩下的鱼数量: "+competitionPond.getFishAmount()+"条");
    }
}
```

数字黑洞

31.1 场景故事

Kaprekar数字黑洞由印度数学家Kaprekar于1949年提出，其提出的操作数字的算法被后人称为Kaprekar算法。对于4位数（从个位到千位的数字分别为a、b、c、d，并且4个数字互不相同），Kaprekar算法如下：

步骤1：把number的各个位上的数字a、b、c、d按照从大到小的顺序排列得到numberOne，再把数字a、b、c、d按照从小到大的顺序排列得到numberTwo，用numberOne−numberTwo得到nextNumber。如果nextNumber等于number，进行步骤3，否则进行步骤2。

步骤2：对新得到的数nextNumber重复上述步骤1操作。

步骤3：结束。

数学上已经证明：对于任何一个4位数，7步以内必然会得到数：6174。

例如，对于：2019

9210−0129=9081

9810−0189=9621

9621−1269=8352

8532−2358=6174

7641−1467=6174

计算4次得到“数字黑洞”——6174。这个6174就称为Kaprekar常数，也称为Kaprekar数字黑洞。对于3位数，按照Kaprekar算法，6步以内必然会得到495。495也是一个Kaprekar数字黑洞，如图31.1所示。

图31.1 Kaprekar数字黑洞

31.2 场景故事的目的

1. 侧重点

结合GUI程序设计巩固Java的常用实用类以及和GUI有关的线程本知识。

（1）GUI事件。

Java的事件处理是基于授权模式，即事件源调用方法将某个对象注册为自己的监视器。Java语言使用接口回调技术实现处理事件的过程，在Java中能触发事件的对象，都用方法：

```
addXXXListener(XXXListener listener)
```

将某个对象注册为自己的监视器，方法中的参数是一个接口。接口listener可以引用任何实现了该接口的类所创建的对象。当事件源触发事件时，接口listener立刻回调类实现的接口中的某个方法。从方法绑定角度来看，Java运行系统要求监视器必须通过绑定某些方法才能处理事件。这就需要用接口来满足此要求，具体操作是将某种事件的处理绑定到对应的接口，即绑定到接口中的方法，也就是说，当事件源触发事件发生后，监视器能够准确地知道调用哪种方法（自动去调用的）。尽可能地让监视器和事件源保持一种松耦合关系，即尽量让事件源所在的类和监视器是组合关系。当事件源触发事件发生后，系统就会知道某个方法将被执行，而无须关心到底是哪个对象调用了这个方法，因为任何实现接口的类的实例都可以作为监视器。

（2）GUI中的线程。

当Java程序包含图形用户界面（GUI）时，Java虚拟机在运行应用程序时会自动启动更多的线程，其中有两个重要的线程：AWT-EventQuecue和AWT-Windows。

AWT-EventQuecue线程负责处理GUI事件，AWT-Windows线程负责将窗体或组件绘制到桌面。JVM要保证各个线程都有使用CPU资源的机会。如果在处理GUI事件时，需要进行复杂或耗时的操作，就应该单独再启动其他用户线程。当把一个线程委派给一个组件事件时要格外小心，比如单击按钮、让线程开始运行，那么当这个线程在执行完run()方法之前，客户可能会随时再次单击该按钮，这时就会发生ILLegalThreadStateException 异常。线程处于“新建”状态时，线程调用isAlive()方法返回false。当一个线程调用start()方法，并占有CUP资源后，该线程的run()方法就开始运行，在线程的run()方法结束之前，即没有进入死亡状态之前，线程调用isAlive()方法返回true。当线程进入“死亡”状态后（实体内存被释放），线程仍可以调用isAlive()方法，这时返回的值是false。在处理GUI事件事要注意让线程调用isAlive()方法，判断线程是否还有实体，如果线程是死亡状态就再分配实体给线程。

2. 涉及的其他知识点

确认对话框是有模式对话框，JOptionPane类的静态方法：

```
public static int showConfirmDialog(Component parentComponent,
Object message,String title,int optionType)
```

得到一个确认对话框，其中参数parentComponent用于指定确认对话框可见时的位置，确认对话框在参数parentComponen指定的组件的正前方显示出来。如果parentComponent为null，则确认对话框会在屏幕的正前方显示出来。Message用于指定对话框上显示的消息；title用于指定确认对话框的标题；optionType可取的有效值是JOptionPane中的类常量：

yes_no_option

yes_no_cancel_option

ok_cancel_option

这些值可以给出确认对话框的外观。

3. 进一步的尝试

如果用户超过3次输入错误，则程序退出执行。

31.3 程序运行效果与视频讲解

视频讲解

在文本框里输入7893，在弹出的对话框里可以看到计算过程，主类是MainClass。程序运行效果如图31.2所示。

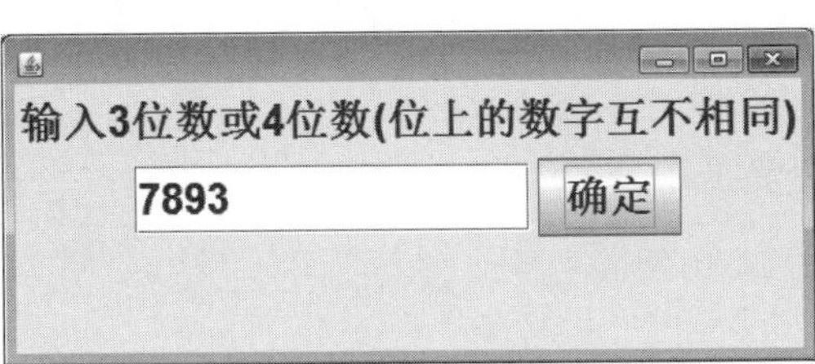

(a) 输入4位数

7893
第1次计算得到6084
第2次计算得到8172
第3次计算得到7443
第4次计算得到3996
第5次计算得到6264
第6次计算得到4176
第7次计算得到6174
第8次计算得到6174
计算了7次得到数字黑洞6174

(b) 显示计算过程

图31.2 GUI程序计算数字黑洞

31.4 阅读源代码

（1）Verification.java的代码如下：

```
import java.util.HashSet;
public class Verification {
    //判断是否是3或4位数(而且每位上的数字互不相同)
    public static boolean verification(long number){
        boolean isWantDigit = true;
        String regex ="[0-9]{3,4}" ;         //匹配3或4位整数
        String str =""+number;
        if(!str.matches(regex)){
            isWantDigit=false;
            return isWantDigit;
```

```
            }
            return judgedDiffrentEachOther(number);
        }
        //判断数字序列含有的数字互不相同
        public static boolean judgedDiffrentEachOther(long number){
            boolean isDifferent = true;
            String numberString =""+number;
          //集合里不会有相同的元素
            HashSet<Character>set = new HashSet<Character>();
            for(char c:numberString.toCharArray()){
                set.add(c);          //数据装箱
            }
            if(set.size()<numberString.length()){
                                       //numberString的字符序列中有相同的
               isDifferent = false;
            }
            return isDifferent;
        }
}
```

（2）ComputerKaprekar.java的代码如下：

```
import java.util.Arrays;
public class ComputerKaprekar {
    public static long getKaprekarSub(long number){
        String str =""+number;
        char [] cTwo =str.toCharArray();
        Arrays.sort(cTwo);
        char [] cOne = new char[cTwo.length];
        for(int i=0;i<cOne.length;i++){ //将cOne中数字按从大到小的顺序排序
           cOne[i] = cTwo[cOne.length-1-i];
        }
        long numberOne = Long.parseLong(new String(cOne));
        long numberTwo = Long.parseLong(new String(cTwo));
        return numberOne- numberTwo;  //返回Kaprekar差
    }
}
```

（3）WindowActionEvent.java的代码如下：

```
import java.awt.*;
import javax.swing.*;
public class WindowActionEvent extends JFrame {
   JTextField inputNumber;
```

```
    JButton button;
    HandleKaprekar listener;
    public WindowActionEvent() {
       init();
       setVisible(true);
       setBounds(10,20,500,200);
       setDefaultCloseOperation(JFrame.EXIT_ON_CLOSE);
    }
    void init() {
       setLayout(new FlowLayout());
       JLabel remind = new JLabel("输入3位数或4位数(位上的数字互不相同)");
       inputNumber = new JTextField(10);
       button = new JButton("确定");
       button.setFont(new Font("",Font.BOLD,26));
       inputNumber.setFont(new Font("",Font.BOLD,26));
       remind.setFont(new Font("",Font.BOLD,26));
       add(remind);
       add(inputNumber);
       add(button);
       listener = new HandleKaprekar();
       listener.setView(this);
       button.addActionListener(listener);
    }
}
```

（4）HandleKaprekar.java的代码如下：

```
import java.awt.event.ActionListener;
import java.awt.event.ActionEvent;
import javax.swing.JOptionPane;
public class HandleKaprekar implements ActionListener {
    Target target;
    WindowActionEvent view;
    Thread thread;          //单独负责计算Kaprekar数的线程
    HandleKaprekar(){
       target = new Target();
       thread = new Thread(target);
    }
    public void setView(WindowActionEvent view) {
        this.view = view;
    }
    public void actionPerformed(ActionEvent e){
        String str=view.inputNumber.getText();
        long number = 0;
```

```
        try{
           number = Long.parseLong(str);
        }
        catch(NumberFormatException exp){
           JOptionPane.showConfirmDialog
           (null,"请输入数字","提示",JOptionPane.WARNING_MESSAGE);
           return;
        }
        if(!Verification.verification(number)){
            JOptionPane.showConfirmDialog
            (null,"输入的数据不符合要求","提示",JOptionPane.WARNING_
            MESSAGE);
            return;
        }
        target.setVisible(true);
        if(!thread.isAlive()){
           thread = new Thread(target);
        }
        try {
           target.setNumber(number);
           thread.start();
        }
        catch(Exception exp){}
    }
}
```

（5）Target.java的代码如下：

```
import java.awt.*;
import javax.swing.*;
public class Target extends JDialog implements Runnable {
    long number;
    JTextArea textShow;
    Target(){
        textShow = new JTextArea(9,80);
        textShow.setLineWrap(true);
        textShow.setFont(new Font("",Font.BOLD,26));
        add(new JScrollPane(textShow));
        setDefaultCloseOperation(JFrame.DISPOSE_ON_CLOSE);
        setBounds(520,20,500,600);
    }
    public void setNumber(long number){
        this.number = number;
        textShow.setText(null);
```

```
        setTitle(""+number);
    }
    public void run() {
        long nextNumber =
        ComputerKaprekar.getKaprekarSub(number);
        int count = 1;
        textShow.append("\n第"+count+"次计算得到"+nextNumber);
        while(nextNumber != number){
           count ++;
           try{
              Thread.sleep(350);
           }
           catch(InterruptedException exp){}
           number = nextNumber;
           nextNumber =ComputerKaprekar.getKaprekarSub(number);
           textShow.append("\n第"+count+"次计算得到"+nextNumber);
        }
        count--;
        textShow.append("\n计算了"+count+"次得到数字黑洞"+number);
    }
}
```

（6）MainClass.java的代码如下：

```
public class MainClass{
   public static void main(String args[ ]) {
       new WindowActionEvent();
   }
}
```

学新概念英语

32.1 场景故事

Tom小时候去过西班牙，和居住在西班牙的表哥相处得很好，而且那时表哥也教会Tom一些西班牙语的单词和句子。最近Tom的表哥从西班牙回国来看望Tom，但表哥的英语不太好，Tom也就经常教表哥一些常用的英语。

一天，Tom想考一考表哥，就拿出一些单词作为考试内容。当考到stupid这一单词时，表哥摇摇头表示不知道是什么意思。

Tom用西班牙小声提示说："笨蛋。"表哥仍然迷惑地摇摇头。

Tom大声吼道："笨蛋！"Tom教表哥英语如图32.1所示。

图32.1 Tom教表哥英语

表哥用西班牙回答："Tom，我真的那么笨吗？"

Tom哈哈大笑地说："我不是说你是笨蛋，而是说stupid这个英文单词的意思是笨蛋！"表哥也破涕为笑。

过了一段时间，Tom的好友Jerry来了。Jerry对Tom说："这样吧，你把新概念英语朗读一遍，特别是第2册，再把课文录制成音频文件，我给你表哥写个程序，

让他可以边看课文边听音频，这样就可以慢慢地提高他的英语水平了。”

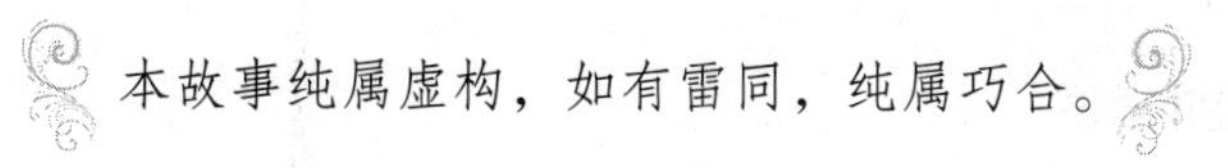

32.2 场景故事的目的

1．侧重点

综合巩固GUI程序设计以及面向对象的基本知识，特别是播放音频的知识。比如，播放声音文件1.wav（必须是Java支持的音频格式，目前不支持MP3格式）。Clip和AudioSystem、AudioInputStream类在javax.sound.sampled包中，步骤如下。

（1）得到Clip对象。

```
Clip clip = AudioSystem.getClip();
```

（2）clip打开音频流。打开到音频文件的音频输入流：

```
File voiceFile = new File("1.wav");
AudioInputStream stream=
AudioSystem.getAudioInputStream(voiceFile);    //得到音频流
clip.open(stream);                             //打开音频流
```

（3）播放或暂停。

```
clip.start();         //开始播放（只播放一次，播放完毕，音频流自动关闭）
clip.loop(int count); //循环播放音频流count+1次（count为负时，无限循环播放）
clip.stop();          //暂停播放
```

（4）关闭音频流。

```
clip.close();
```

一旦关闭，clip就不能再播放（调用start方法无效），除非clip能重新打开音频流。

Clip对象是个守护线程，即当Clip对象播放音频时，程序仍然可以做其他的事情（这类似于编辑Word文档时选择了打印，并不影响继续编辑Word文档）。

2．涉及的其他知识点

FileReader流是Reader的子类对象调用int read（char b[], int off, int len）方法从源中试图读取len个字符到字符数组b中，并返回实际读取的字符数目。如果到达文件的末尾，则返回-1，参数off用于指定从字符数组的某个位置开始存放读取的数据。

3．进一步的尝试

增加一个JLabel标签，用于显示用户选择的课文的标题。

32.3 程序运行效果与视频讲解

视频讲解

需要在程序所在的当前目录下建立名字是“音频文件”和“课文”的文件夹。将课文和对应的音频文件各自放在两个文件夹里。程序运行后，单击课文下拉列表，选择一篇课文，然后单击“播放”按钮。主类是MainClass。程序运行效果如图32.2所示。

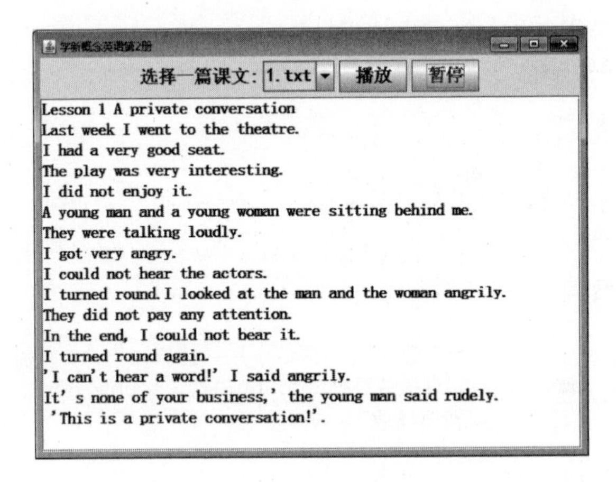

图32.2　学新概念英语的程序运行效果

32.4 阅读源代码

（1）ReadWindow.java的代码如下：

```
import javax.swing.*;
import java.awt.*;
import java.io.*;
import javax.sound.sampled.Clip;
import javax.sound.sampled.AudioSystem;
```

```
public class ReadWindow extends JFrame{
    JButton playButton,stopButton;
    //存放课文文件的名字，例如第一课的名字1.txt
    JComboBox<String> listFile ;
    JTextArea textShowEnglish;
    //负责处理单击按钮触发的ActionEevnt
    HandleButton  handleButton;
    //负责处理选中下拉列表选项触发的ActionEevnt
    HandleComboBox handleComboBox;
    Clip clip;                                        //音频对象（具有守护线程特点）
    ReadWindow(){
        setTitle("学新概念英语第2册");
        initWindow();                                 //初始窗口中的组件
        initListener();                               //为按钮和下拉列表注册监视器
        validate();
        setBounds(10,20,1200,650);
        setDefaultCloseOperation(JFrame.DISPOSE_ON_CLOSE);
        setVisible(true);
    }
    public void initWindow() {
        playButton = new JButton("播放");
        stopButton = new JButton("暂停");
        listFile = new JComboBox<String>();
        JPanel pNorth = new JPanel();
        JLabel mess = new JLabel("选择一篇课文:");
        pNorth.add(mess);
        pNorth.add(listFile);
        pNorth.add(playButton);
        pNorth.add(stopButton);
        initList();                                   //初始化下拉列表中的选项
        textShowEnglish=new JTextArea();
        playButton.setFont(new Font("宋体",Font.BOLD,20));
        stopButton.setFont(new Font("宋体",Font.BOLD,20));
        listFile.setFont(new Font("黑体",Font.BOLD,20));
        mess.setForeground(new Color(0,63,125));
        mess.setFont(new Font("宋体",Font.BOLD,20));
        playButton.setForeground(Color.blue);
        stopButton.setForeground(Color.red);
        textShowEnglish.setFont(new Font("宋体",Font.BOLD,18));
        textShowEnglish.setForeground(new Color(0,60,0));
        textShowEnglish.setBackground(Color.white);
        add(new JScrollPane(textShowEnglish),BorderLayout.CENTER);
        add(pNorth,BorderLayout.NORTH);
        validate();
```

```
    }
    public void initList(){                         //初始化下拉列表中的选项
         File dirFile = new File("课文");
         String fileName[] = dirFile.list();
         for(String name:fileName) {
             listFile.addItem(name);
         }
    }
    public void initListener(){                     //为按钮下拉列表注册监视器
        handleButton = new HandleButton();
        handleComboBox = new HandleComboBox();
       //将当前窗口（引用）传递个handleButton的view
        handleButton.setView(this);
        handleComboBox.setView(this);
        playButton.addActionListener(handleButton);
        //注册监视器
        stopButton.addActionListener(handleButton);
        listFile.addActionListener(handleComboBox);
    }
}
```

（2）ReadWindow.java的代码如下：

```
import java.awt.event.*;
import java.io.*;
import javax.sound.sampled.AudioSystem;
import javax.sound.sampled.AudioInputStream;
public class HandleComboBox implements ActionListener {
     ReadWindow view;
     public void setView(ReadWindow view) {
          this.view = view;
     }
     public void actionPerformed(ActionEvent e){
        try{
           if(view.clip!=null){
               view.clip.close();          //关闭正在播放或暂停的音频流
           }
           view.clip = AudioSystem.getClip();//得到Clip对象，负责播放音频文件
        }
        catch(Exception exp){}
        String fileName = view.listFile.getSelectedItem().toString();
        readingFile(fileName);                      //将课文读入文本区
       String voiceName = fileName.substring(0,fileName.lastIndexOf("."));
          try {
```

```
            File voiceFile = new File("音频文件/"+voiceName+".wav");
            System.out.println(voiceFile.getName());
                                                //也可以不输出名字
            AudioInputStream stream=
            AudioSystem.getAudioInputStream(voiceFile);
                                                //得到音频流
            view.clip.open(stream);             //打开音频流
        }
        catch(Exception ee){
            System.out.println(ee);
        }
    }
    void readingFile(String fileName) {         //将课文读入文本区
        view.textShowEnglish.setText(null);
        int n=-1;
        char [] a=new char[100];
        try{  File f=new File("课文/"+fileName);
            Reader in = new FileReader(f);
            while((n=in.read(a,0,100))!=-1) {
                String s=new String (a,0,n);
                view.textShowEnglish.append(s);
            }
            in.close();
        }
        catch(IOException e) {
            view.textShowEnglish.setText("File read Error"+e);
        }
    }
}
```

（3）HandleButton.java的代码如下：

```
import java.awt.event.*;
import java.io.*;
public class HandleButton implements ActionListener {
    ReadWindow view;
    public void setView(ReadWindow view) {
        this.view = view;
    }
    public void actionPerformed(ActionEvent e) {
        if(e.getSource() == view.playButton) {
            view.clip.start();    //开始播放音频流（只播放一次）
            view.clip.loop(1);    //循环播放音频流一次（即一共播放两次）
        }
```

```
            else if(e.getSource() == view.stopButton) {
               view.clip.stop();      //暂停播放音频流
            }
      }
}
```

（4）MainClass.java的代码如下：

```
public class MainClass {
    public static void main(String args[]){
         new ReadWindow();
    }
}
```

老鼠走迷宫

33.1 场景故事

Tom自从和Java村的女孩结婚后，也迷恋上了Java编程。时常说些类、接口、面向抽象、面向接口编程之类的话。一天，Tom的夫人让他编写一个“老鼠走迷宫”的程序，并跟他说了一些要求：迷宫是由m×n个格子组成的矩形，有些格子代表路，有些格子代表墙；要求通过读写一个文本文件来设置迷宫，即设置哪些格子是路、哪些格子是墙；然后让老鼠从入口进入迷宫，并从出口走出迷宫。

Tom绞尽脑汁地想了半天，不知该如何操作，特别是关于老鼠走迷宫的算法一直都没有什么想法。Tom决定向Jerry请教。

Jerry对Tom说：“我可以告诉你一个秘籍。一般情况下，老鼠走迷宫，看见可走的路就走，直到无路可走时再向后退，秘籍在于必须记住来时的路，一直后退到某个路口，发现该路口周围还有未曾走过的路，即不曾走过的岔路口，然后随便挑选一个岔路口走下去。重复这个办法，只要迷宫中的路有一条是通向出口的，那么老鼠就一定能走出去，否则最后也会回到入口，而不会迷失在迷宫里。比如，当老鼠走了5步（1、2、3、4、5），在第5步发现除了来时的路已经无路可走。幸运的是，你事先记住了这5步，然后可以按照“后记先退”的原则，即从第5步开始往后退，假如退到3，发现在第3步这个地方还有没走过的岔路口，就不再退了，在这个地方随便找个岔路口继续走下去。”Jerry不慌不忙地喝了口茶接着说：“还有个小秘密，那就是曾走过的路尽管是用来的往后退的，但在你心里要把它们当墙看，这样你会节省很多时间。”Jerry给Tom传授走迷宫秘籍如图33.1所示。

Tom茅塞顿开，发现可以用Java中的Sack（堆栈）对象记住走过的路，Sack对象的数据结构的特点就是“先进后出”，也就是Jerry说的“后记先退”啊。很快，Tom就完成了夫人的任务，得到好评。

本故事纯属虚构，如有雷同，纯属巧合。

图33.1　Jerry给Tom传授走迷宫秘籍

33.2　场景故事的目的

1. 侧重点

巩固面向抽象、面向接口以及MVC的设计思想，巩固线程的有关知识。

2. 涉及的其他知识点

javac.swing包中的部分类，如JFame、JButton、JLabel等。

3. 进一步的尝试

（1）输出老鼠走过的方块数目，包括重复走过的方块。
（2）用某种算法随机生成一个迷宫。

33.3　程序运行效果与视频讲解

视频讲解

Tom用下列jerry.txt文件生成迷宫。主类是MainClass。程序运行效果如图33.2所示。

jerry.txt文件：

*11001111000111111111000

001110110000011001111000
000111110000111010111111
111010111111111110011110
110110011010001111110000
101010111111111010111111
111110011110001101110001
001111010000001100110001
11101011111100111111000#
111110011111001111110000

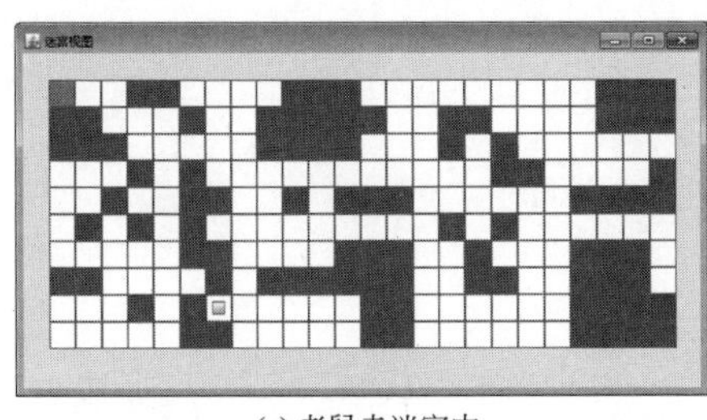

(a) 老鼠走迷宫中

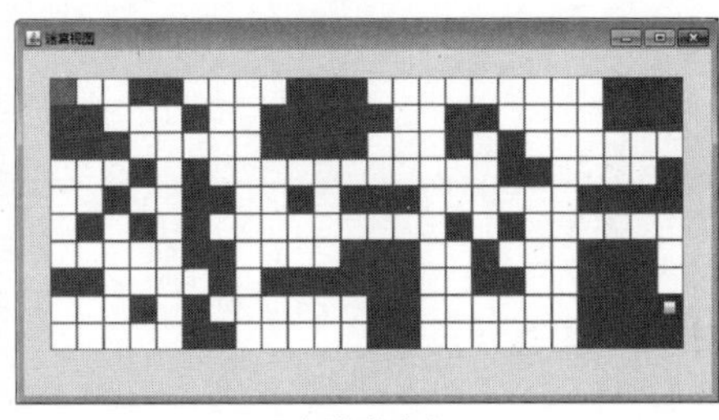

(b) 老鼠到达出口

图33.2 老鼠走迷宫

数据模型的类的包名是mouse.data，视图相关的类的包名是mouse.view。将数据模型和视图相关的类按照包名对应的路径保存，保持路径对齐的原则，即mouse对应的父目录相同，比如tom目录。用命令行进入tom目录，如下编译：

```
tom> javac mouse\data\*.java
tom> javac mouse\view\*.java
```

运行如下代码：

```
tom> java mouse.data.MainClass
```

另外，需要将名字是jerry.txt的文件（用于生成迷宫）保存在和当前应用程序相同的目录中，即和mouse同目录同级别的目录，如tom目录。

33.4 阅读源代码

（1）Maze.java的代码如下：

```
package mouse.data;
public abstract class Maze {
     public char point[][];                       //point[i][j]是迷宫中的点
     public int startI,startJ;                    //迷宫入口的索引
     int row = 0;                                 //迷宫行数
     public int getStartI(){
        return startI;
     }
     public int getStartJ(){
        return startJ;
     }
     public char[][] getMaze(){
        return point;
     }
     public int getRow(){
        return row;
     }
}
```

（2）View.java的代码如下：

```
package mouse.view;
public interface View {
    public void setPosition(int i,int j);
    public void setPoint(char a[][]);
}
```

（3）MazeByFile.java的代码如下：

```
package mouse.data;
import java.io.File;
import java.io.RandomAccessFile;
import java.io.IOException;
import java.util.Arrays;
import java.util.HashSet;
public class MazeByFile extends Maze{
     File mazeFile;                       //初始化迷宫的用的文件
     HashSet<Integer> set;                //检查二维数组中的一维数组是否等长
     public  MazeByFile(File f){
         set = new HashSet<Integer>();
         mazeFile = f;
         initMaze();
     }
     private void initMaze() {
```

```
        RandomAccessFile in=null;
        row = 0;
        try{
          in=new RandomAccessFile(mazeFile,"r");
          long length=in.length();
          long position=0;
          in.seek(position);
          while(position<length){
             String str=in.readLine().trim();
             set.add(str.length());
             if(set.size()>1) {
                 System.out.println("迷宫不合理");
                 System.exit(0);                    //退出程序
             }
             position=in.getFilePointer();
             row++;
          }
          point=new char[row][];                    //首先创建二维数组
          position=0;
          in.seek(position);                        //再定位到文件的开头
          length=in.length();
          int i = 0;
          while(position<length){
             String str = in.readLine().trim();
             point[i] = new char[str.length()];//再创建二维数组中的一维数组
             for(int j=0;j<str.length();j++){
                  point[i][j] = str.charAt(j);
                  if(point[i][j]=='*') {
                    startI =i;
                    startJ =j;
                  }
             }
             position=in.getFilePointer();
             i++;
          }
      }
      catch(IOException exp){}
   }
}
```

（4）Person.java的代码如下：

```
package mouse.data;
import mouse.view.*;
```

```
import java.util.Stack;
import java.util.Arrays;
import java.util.EmptyStackException;
public class Person extends Thread{
     Maze maze;
     char [][] a;                                        //迷宫
     public View viewPerson;                             //走迷宫者的外观
     public void setMaze(Maze maze){
         this.maze = maze;
         a = maze.getMaze();

     }
     public void setView(View view){
        viewPerson = view;
        viewPerson.setPoint(a);
     }
     private boolean alreadyNoWay(int i,int j){
                                                         //判断i、j周围是否无路可走
         int row = maze.getRow();
         int column = a[i].length;
         boolean isNoway = true;
         if(i<row-1){
             isNoway = isNoway&&a[i+1][j]=='0';
         }
         if(i>1){
             isNoway = isNoway&&a[i-1][j]=='0';
         }
         if(j<column-1){
             isNoway = isNoway&&a[i][j+1]=='0';
         }
         if(j>1){
             isNoway = isNoway&&a[i][j-1]=='0';
         }
         return isNoway;
     }
     public void run(){
         try{
            walkMaze();
         }
         catch(EmptyStackException exp){
            System.out.println("无出口");
            System.out.println("一定回到了入口");
         }
     }
```

```
private void walkMaze() throws EmptyStackException {
    Stack <Integer> saveI = new Stack <Integer>();
    //存放走过的路的索引坐标i
    Stack <Integer> saveJ = new Stack <Integer>();
    //存放走过的路的索引坐标j
    int i =maze.getStartI();                    //起点
    int j = maze.getStartJ();
    a[i][j] = '0';
    viewPerson.setPosition(i,j);                //视图达到指定地点
    while(true){
       saveI.push(i);                           //记下走过的路的位置(压栈)
       saveJ.push(j);                           //记下走过的路的位置
       boolean boo = alreadyNoWay(i,j);
       while(boo){
           i= saveI.pop();
           j = saveJ.pop();                     //退回一步
           try {
              sleep(500);                       //休息500ms回退一步
           }
           catch(InterruptedException exp){}
           viewPerson.setPosition(i,j);  //视图达到指定地点
           boo = alreadyNoWay(i,j);
           //一直退到a[i][j]的周围有路可走
       }
       int m =0,n=0;
       int row = maze.getRow();
       int column = a[i].length;
       if(i<row-1){  //检查a[i][j]的南面是否有没走过的路
          if(a[i+1][j]=='1'||a[i+1][j]=='#'){
             m=i+1;
             n =j;
          }
       }
       if(i>0){
          if(a[i-1][j]=='1'||a[i-1][j]=='#'){  //北
            m=i-1;
            n =j;
          }
       }
       if(j<column-1){
          if(a[i][j+1]=='1'||a[i][j+1]=='#'){  //东
             m=i;
             n =j+1;
          }
```

```
            }
            if(j>0){
               if(a[i][j-1]=='1'||a[i][j-1]=='#'){  //西
                 m=i;
                 n =j-1;
               }
            }
            if(a[m][n] =='#'){
                System.out.println(a[m][n]+"出口");
                viewPerson.setPosition(m,n);  //视图达到指定地点
                break;
            }
            a[m][n] = '0';      //设置成'0'表示走过该点了
            i = m;              //然后还要压栈，为了允许后退到该点
            j = n;
            viewPerson.setPosition(i,j);  //视图达到指定地点
            try {
                sleep(500);                    //休息500ms继续找下一个路点
            }
            catch(InterruptedException exp){}
        }
    }
}
```

（5）MouseView.java的代码如下：

```
package mouse.view;
import javax.swing.JButton;
public class MouseView extends JButton implements View {
    GUIView guiView;
    public char a[][];
    int x;                    //走迷宫者位置
    int y;
    int width;
    int height;
    int offset;
    public void setPoint(char a[][]){
        this.a = a;
        guiView = new GUIView();
        guiView.add(this);
        guiView.setPoint(a);
        width = guiView.width;
        height = guiView.height;
        offset = guiView.offset;
```

```
        guiView.setVisible(true);
    }
    public void setPosition(int i,int j){
       x =j*width+offset+width/4;
       y= i*height+offset+height/4;
       this.setBounds(x,y,width/2,height/2);
    }
}
```

（6）GUIView.java的代码如下：

```
package mouse.view;
import java.awt.geom.*;
import javax.swing.*;
import java.awt.*;
public class GUIView extends JFrame {
    char a[][];
    int width = 26;                              //绘制的小矩形的宽
    int height = 26;
    int offset =26;
    JButton geometry[][];                        //迷宫的外观
    public GUIView(){
       setTitle("迷宫视图");
       setLayout(null);
       setBounds(20,20,760,500);
       setDefaultCloseOperation(JFrame.EXIT_ON_CLOSE);
       setVisible(true);
    }
    public void setPoint(char a[][]){
        this.a = a;         //将迷宫二维数组传递窗口，以便绘制迷宫图
        geometry = new JButton[a.length][a[0].length];
        for(int i=0;i<geometry.length;i++){
           for(int j=0;j<geometry[i].length;j++){
              geometry[i][j] = new JButton();
              if( a[i][j]=='1'){
                 geometry[i][j].setBackground(Color.white);
              }
              else if( a[i][j] =='0'){
                 geometry[i][j].setBackground(Color.gray);
              }
              else if( a[i][j] =='*'){
                 geometry[i][j].setBackground(Color.green);
              }
              else if( a[i][j] =='#'){
```

```
                    geometry[i][j].setBackground(Color.red);
                }
                add(geometry[i][j]);
                geometry[i][j].setBounds(j*width+offset,i*height+offset,
                width,height);
                geometry[i][j].repaint();
            }
        }
        validate();
        repaint();
    }
}
```

（7）MainClass.java的代码如下：

```
package mouse.data;
import mouse.view.*;
import java.io.File;
public class MainClass {
    public static void main(String args[]){
        Person tom  = new Person();
        Maze maze = new MazeByFile(new File("jerry.txt"));
        tom.setMaze(maze);
        View view = new MouseView();
        tom.setView(view);
        tom.start();
    }
}
```

34.1 场景故事

生命游戏是英国数学家约翰·何顿·康威于1970年发明的细胞自动机。生命游戏属于二维细胞自动机的一种（又称为二维元胞自动机）。每一个格子都可以看成是一个生命体，每个生命都有“生和死”两种状态，每一个格子旁边都有邻居格子存在，如果把由3×3的9个格子构成的正方形看作一个基本单位，那么这个正方形中心的格子的邻居就是它旁边的8个格子（至多8个）。生命游戏规则如图34.1所示。

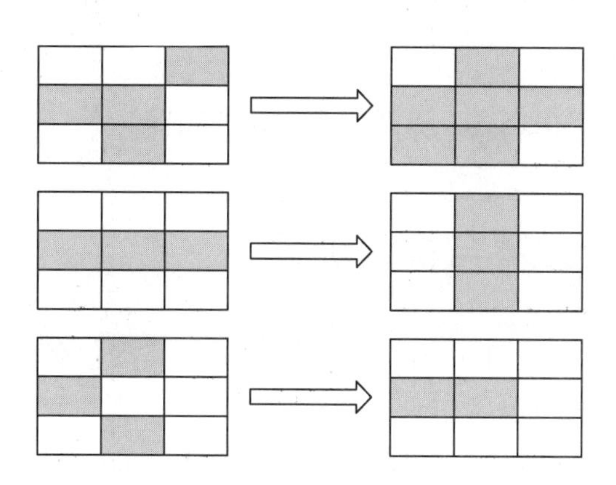

图34.1 生命游戏规则

（1）如果一个细胞周围有3个细胞为生，则该细胞为生（一个细胞周围至多有8个细胞；如果当前细胞原本为死，则转为生；如果当前细胞原本为生，则保持不变）。

（2）如果一个细胞的周围有两个细胞为生，则该细胞的生死状态保持不变。

（3）在其他情况下，该细胞为死（即该细胞若原本为生，则转为死；如果该细胞若原本为死，则保持不变）。

最早研究细胞自动机的科学家是冯·诺依曼（被誉为计算机之父）。后来英国数学家约翰·何顿·康威给出了有趣的二维细胞自动机程序——Life Game。在Life Game中（有限维），给出任何初始状态，运行一段时间后，元胞空间可能趋于一个空间平稳的构形，称进入平稳状态，即每一个元胞处于固定状态，不随时间变化而变化；但有时也会进入一个周期状态，即在几个状态中周而复始。

34.2 场景故事的目的

1. 侧重点

生命游戏的算法。

2. 涉及的其他知识点

GUI程序设计和MVC的设计理念以及线程和读取文件的知识。

3. 进一步的尝试

生命游戏进入稳定状态后，更换代表生命的图像。

34.3 程序运行效果与视频讲解

视频讲解

数据模型的类的包名是life.game，视图相关的类的包名是life.view。将数据模型和视图相关的类按照与包名相对应的路径保存，并保持路径对齐的原则，即life类对应的父目录相同，比如moon目录。用命令行进入moon目录，如下编译：

```
moon> javac life\game\*.java
moon> javac life\view\*.java
```

运行（主类是life.view.LifeGame）：

```
moon> java life.view.LifeGame
```

为了让用户可以更加方便地更改游戏的格子数目，当程序运行时需要读取名字为“宽高配置.txt”文件。该文件和当前应用程序在相同的目录中，即和life目录同级别的目录中，比如moon目录中。“宽高配置.txt”文件内容由用户维护和修改：

高26
宽55
格子大小22

另外，需要将名字为shui.jpg的图像文件（显示生命活的样子）保存在当前应用程序在相同的目录中，即和life目录同级别的目录中，如moon目录中。程序运行后，用户可设置初始状态，初始状态如图34.2（a）所示，然后生命开始进行演变，演变过程如图34.2（b）所示，最后趋于一个稳定的状态，如图34.2（c）所示。

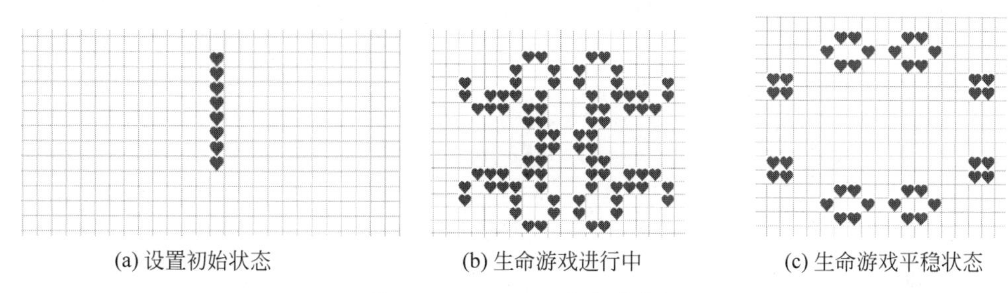
(a) 设置初始状态　(b) 生命游戏进行中　(c) 生命游戏平稳状态

图34.2　Life Game生命游戏

34.4 阅读源代码

（1）BoxPoint.java的代码如下：

```
package life.game;
public class BoxPoint{
    int x,y;
    boolean isAlive;
    public void setX(int x){
        this.x=x;
    }
    public void setY(int y){
        this.y=y;
    }
    public int getX(){
        return x;
    }
    public int getY(){
        return y;
    }
    public void setIsAlive(boolean isAlive){
        this.isAlive = isAlive;
```

```
    }
    public boolean isAlive(){
        return isAlive;
    }
}
```

（2）LifeArea.java的代码如下：

```
package life.game;
import java.util.*;
public class LifeArea {
    public BoxPoint [][] point;                              //刻画生命状态的点
    HashSet<BoxPoint>set = new HashSet<BoxPoint>();    //存放周围8个点
    public void setAllBoxPoint(BoxPoint[][] point){
        this.point = point;
    }
    public void clearLife(){
        for(int i=0;i<point.length;i++){
          for(int j=0;j<point[i].length;j++){
             point[i][j].setIsAlive(false);
          }
        }
    }
    public BoxPoint [][] getNextAllBoxPoint(){          //返回下一代
       BoxPoint [][] p = copyPoint(point);
       for(int m=0;m<p.length;m++){
           for(int n=0;n<p[m].length;n++){              //Life Game的规则
                int count = backLifeIsAliveAround(m,n,p);
                //返回周围生命的个数
                if(count == 3){
                    point[m][n].setIsAlive(true);
                }
                else if(count == 2){
                    point[m][n].setIsAlive(point[m][n].isAlive());
                }
                else {
                    point[m][n].setIsAlive(false);
                }
           }
       }
       return point;
    }
    public int backLifeIsAliveAround(int m,int n,BoxPoint [][] p){
        int count = 0;
```

```
        set.clear();
        if(m<p.length-1)                                  //下
           set.add(p[m+1][n]);
        if(m>=1)
           set.add(p[m-1][n]);                            //上
        if(n<p[0].length-1)
           set.add(p[m][n+1]);                            //右
        if(n>=1)
           set.add(p[m][n-1]);                            //左
        if(m<p.length-1&&n<p[0].length-1)
            set.add(p[m+1][n+1]);                         //下右
        if(m<p.length-1&&n>=1)
            set.add(p[m+1][n-1]);                         //下左
        if(m>=1&&n>=1)
            set.add(p[m-1][n-1]);                         //上左
        if(m>=1&&n<p[0].length-1)                         //上右
            set.add(p[m-1][n+1]);
        Iterator<BoxPoint> iter=set.iterator();
        while(iter.hasNext()){
           BoxPoint te = iter.next();
           if(te.isAlive())
              count++;
        }
        return count;
    }
    BoxPoint [][] copyPoint(BoxPoint [][] p){
          BoxPoint [][] point = new BoxPoint[p.length][p[0].length];
          for(int i=0;i<point.length;i++){
             for(int j=0;j<point[i].length;j++){
                 point[i][j] = new BoxPoint();
                 point[i][j].setIsAlive(p[i][j].isAlive());
             }
          }
          return point;
    }
}
```

（3）Contant.java的代码如下：

```
package life.game;
public class Contant {
   public static int WIDTH = 22;
   public static int HEIGHT = 22;
   public static int LEFT_X = 40;
```

```
    public static int LEFT_Y = 10;
}
```

（4）LifeAreaView.java的代码如下：

```
package life.view;
import life.game.*;
import javax.swing.JPanel;
import javax.swing.Timer;
import javax.swing.JLabel;
import java.awt.event.*;
import java.awt.*;
import java.util.ArrayList;
public class LifeAreaView extends JPanel implements ActionListener{
    public LifeArea lifeArea ;
    public BoxPoint [][] point;
    public int width = Contant.WIDTH;                  //点之间的横向距离
    public int height = Contant.HEIGHT;                //点之间的纵向距离
    public int leftX = Contant.LEFT_X;                 //左上角起点偏移坐标
    public int leftY = Contant.LEFT_Y;
    Timer thread;                                      //演示生命游戏
    HandleInitLife handleInitLife; //负责给出第一代的生命分布（用鼠标事件）
    public JLabel showMess,fixMess;
    public int n;                                      //迭代次数
    Image image;
    Toolkit tool;
    ArrayList<BoxPoint [][]> list;                     //负责判断周期点
    public int daiCycleAppear = -1;     //负责存放出现周期时的迭代次数
    public LifeAreaView(){
        list = new ArrayList<BoxPoint [][]>();
        setBackground(Color.white);
        thread = new Timer(800,this);
        handleInitLife = new HandleInitLife();
        handleInitLife.setView(this);
        //用户单击细胞盒子，设置该盒子是不是有生命
        addMouseListener(handleInitLife);
        addMouseMotionListener(handleInitLife);
        showMess = new JLabel();
        fixMess = new JLabel();
        fixMess.setFont(new Font("",Font.BOLD,28));
        fixMess.setForeground(Color.red);
        showMess.setFont(new Font("",Font.BOLD,28));
        showMess.setText("这是生命游戏的第"+n+"代");
        tool = getToolkit();
```

```
    image = tool.getImage("shui.jpg");
}
public void setLifeArea(LifeArea lifeArea){
     this.lifeArea = lifeArea;
     point = lifeArea.point;
     initPoint();
     repaint();
}
public void setBoxSize(int size) {
   width = height = size;
}
private void initPoint(){

    daiCycleAppear = -1;
    n = 0;
    for(int i=0;i<point.length;i++){
      for(int j=0;j<point[i].length;j++){
      //组件坐标系原点是左上角，向右是x轴，向下是y轴
         point[i][j].setX(j*width+leftX);
         point[i][j].setY(i*height+leftY);
      }
   }
}
public void clearLife(){
   daiCycleAppear = -1;
   n = 0;
   list = new ArrayList<BoxPoint [][]>();
   thread.stop();
   lifeArea.clearLife();
   showMess.setText("这是生命游戏的第"+n+"代");
   fixMess.setText("");
   repaint();
}
public void startLifeGame(){
    thread.start();
}
public void stopLifeGame(){
    thread.stop();
}
public void paintComponent(Graphics g){
  super.paintComponent(g);
  Color c = new Color(200,200,200);
  for(int i=0;i<point.length;i++){
      for(int j=0;j<point[i].length;j++){
```

```
            g.setColor(c);
            g.drawRect(point[i][j].getX(),point[i][j].getY(),width,
            height);
            if(point[i][j].isAlive()) {
                int x = point[i][j].getX();
                int y = point[i][j].getY();
                  //g.setColor(Color.green);也可以不画图像，画个圆圈也行
                   //g.fillOval(x,y,width,height);
                   g.drawImage(image,x,y,width,height,this);
            }
         }
      }
   }
   public void actionPerformed(ActionEvent e) {
         boolean stop = true;
         repaint();
         n++;
         BoxPoint[][]p = copyPoint(point);
         if(list.size()==0){
            list.add(p);
         }
         else {
            if(repeat() == -1)
               list.add(p);
         }
         point = lifeArea.getNextAllBoxPoint();

         showMess.setText("这是生命游戏的第"+n+"代");
         showMess.repaint();
         int m =0;
         if(((m=repeat())!=-1)&&(daiCycleAppear == -1)){
             if(daiCycleAppear == -1)
               daiCycleAppear = n;
             fixMess.
             setText("从第"+daiCycleAppear+"代开始出现"+( list.size()-m)
             +"周期点");
         }
         if(daiCycleAppear !=-1)
            tool.beep();
   }
   BoxPoint [][] copyPoint(BoxPoint [][] p){
          BoxPoint [][] point = new BoxPoint[p.length][p[0].length];
          for(int i=0;i<point.length;i++){
             for(int j=0;j<point[i].length;j++){
```

```
                point[i][j] = new BoxPoint();
                point[i][j].setIsAlive(p[i][j].isAlive());
            }
        }
        return point;
    }
    int repeat(){
        boolean ok = true;          //是否找到周期点
        int index = -1;
        for(int m=0;m<list.size();m++) {
            ok = true;
            BoxPoint[][] point_m = list.get(m);
            for(int i=0;i<point.length;i++){
              for(int j=0;j<point[i].length;j++){
                 if(point[i][j].isAlive() != point_m[i][j].isAlive()){
                     ok = false;
                     break;
                 }
              }
              if(ok == false) break;
            }
            if(ok == true) {
              index = m;
              break;
            }
        }
        return index;
    }
}
```

（5）HandleInitLife.java的代码如下：

```
package life.view;
import java.awt.event.*;
import java.awt.Rectangle;
public class HandleInitLife extends MouseAdapter implements MouseMotionListener
{
    LifeAreaView view;
    public void setView(LifeAreaView view) {
        this.view = view;
    }
    public void mousePressed(MouseEvent e){
       if(view.n == 0){       //按下鼠标左键
          if(e.getButton()==MouseEvent.BUTTON1) {
```

```
                int x = e.getX();
                int y = e.getY();
                setLife(x,y);
            }
            else if(e.getButton()==MouseEvent.BUTTON3) {
                int x = e.getX();
                int y = e.getY();
                cancelLife(x,y);
            }
        }
    }
    public void mouseClicked(MouseEvent e){
        if(view.n == 0){
            if(e.getClickCount()>=2) {
                int x = e.getX();
                int y = e.getY();
                cancelLife(x,y);
            }
        }
    }
    private void setLife(int x,int y){
        view.daiCycleAppear = -1;
        for(int i=0;i<view.point.length;i++){
            for(int j=0;j<view.point[i].length;j++){
                int px = view.point[i][j].getX();
                int py = view.point[i][j].getY();
                Rectangle rect = new Rectangle(px,py,view.width,view.height);
                if(rect.contains(x,y))
                  view.point[i][j].setIsAlive(true);
            }
            view.repaint();
        }
    }
    private void cancelLife(int x,int y){
        view.daiCycleAppear = -1;
        for(int i=0;i<view.point.length;i++){
            for(int j=0;j<view.point[i].length;j++){
                int px = view.point[i][j].getX();
                int py = view.point[i][j].getY();
                Rectangle rect = new Rectangle(px,py,view.width,view.height);
                if(rect.contains(x,y))
                  view.point[i][j].setIsAlive(false);
            }
            view.repaint();
```

```
            }
        }
        public void mouseMoveed(MouseEvent e){}
        public void mouseDragged(MouseEvent e){
            if(view.n == 0) {
                int x = e.getX();
                int y = e.getY();
                setLife(x,y);
            }
        }
}
```

（6）LifeGame.java的代码如下：

```
package life.view;
import life.game.LifeArea;
import life.game.BoxPoint;
import javax.swing.*;
import java.awt.*;
import java.awt.event.*;
import java.util.Scanner;
import java.io.*;
public class LifeGame extends JFrame {        //也是主类
    File file;
    LifeAreaView lifeAreaView;
    JButton initLifeGame,startLifeGame,stopLifeGame;
    int width = 20;
    int height = 50;
    int size = 22;                                  //格子大小（正方形的宽）
    Scanner scanner;
    public LifeGame(){
        file = new File("./宽高配置.txt");
        try {
            scanner = new Scanner(file);
            scanner.useDelimiter("[^0123456789]+");
            width= scanner.nextInt();
            height= scanner.nextInt();
            size = scanner.nextInt();
        }
        catch(IOException exp){  System.out.println(exp);
            width = 20;
            height = 50;
            size = 22;
        }
```

```
LifeArea lifeArea = new LifeArea();
setVisible(true);
setBounds(10,5,1300,726);
setDefaultCloseOperation(JFrame.EXIT_ON_CLOSE);
validate();
initLifeGame = new JButton("重新开始");
startLifeGame = new JButton("开始(继续)游戏");
stopLifeGame = new JButton("暂停游戏");
BoxPoint [][] point = new BoxPoint[width][height];
for(int i=0;i<point.length;i++){
  for(int j=0;j<point[i].length;j++){
     point[i][j] = new BoxPoint();
  }
}
lifeArea.setAllBoxPoint(point);
lifeAreaView = new LifeAreaView();
lifeAreaView.setBoxSize(size);
lifeAreaView.setLifeArea(lifeArea);
JPanel northPanel = new JPanel();
JLabel tips =
new JLabel("在方格上单击并设置生命点,右击取消设置");
northPanel.add(tips);
JPanel southPanel = new JPanel();
southPanel.add(lifeAreaView.fixMess);
southPanel.add(startLifeGame);
southPanel.add(stopLifeGame);
southPanel.add(initLifeGame);
southPanel.add(lifeAreaView.showMess);
add(northPanel,BorderLayout.NORTH);
add(southPanel,BorderLayout.SOUTH);
add(lifeAreaView,BorderLayout.CENTER );
initLifeGame.addActionListener(new ActionListener(){
    public void actionPerformed(ActionEvent e){
       lifeAreaView.fixMess.setText(null);
       lifeAreaView.clearLife();
       lifeAreaView.n = 0;
    }
});
startLifeGame.addActionListener(new ActionListener(){
    public void actionPerformed(ActionEvent e){
       lifeAreaView.startLifeGame();
    }
});
stopLifeGame.addActionListener(new ActionListener(){
```

```
            public void actionPerformed(ActionEvent e){
               lifeAreaView.stopLifeGame();
            }
         });
         lifeAreaView.repaint();
         validate();
      }
      public static void main(String [] args){
         LifeGame lifeGame = new LifeGame();
      }
}
```

场景故事 35

牵　手

35.1　场景故事

从高空俯瞰忙忙碌碌的人们，感觉人就好像沧海一粟，如图35.1所示。很多人经常感叹："众里寻他千百度。蓦然回首，那人却在，灯火阑珊处。"相识满天下，但是真正的知己又有几个呢？真正了解自己和关心自己的朋友又有几个呢？相遇是缘分、还是幸运？"也有很多倾诉这样情感的歌曲，比如曾经流行一时的经典歌曲《牵手》。

图35.1　茫茫人海牵手不易

35.2　场景故事的目的

1. 侧重点

面向抽象、面向接口以及MVC的设计思想。

2. 涉及的其他知识点

线程以及绘制基本图形、图像的知识。

3. 进一步的尝试

在命令行或一个文本区显示牵手的时间日期。

35.3 程序运行效果与视频讲解

视频讲解

数据模型的类的包名是qiansou.data，视图相关的类的包名是qianshou.view。将数据模型和视图相关的类按照包名对应的路径保存，保持路径对齐的原则，即qianshou对应的父目录相同，比如Love目录。用命令行进入Love目录，如下编译：

```
Love> javac qianshou\data\*.java
Love> javac qianshou\view\*.java
```

运行代码如下（主类是WindowView）：

```
Love> java qianshou.view.WindowView
```

程序运行效果如图35.2所示。另外，需要将名字是boy.png和girl.png的文件（用于显示人物的外观）保存在当前应用程序在相同的目录中，即和qianshou同级别的目录中，比如Love目录中。

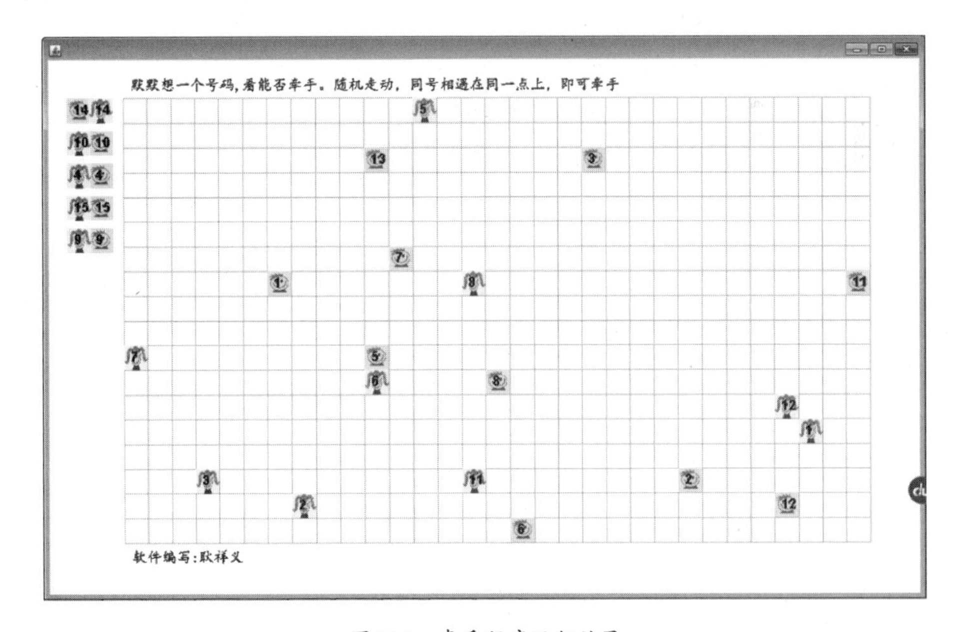

图35.2　牵手程序运行效果

35.4 阅读源代码

（1）Contant.java，系统需要的重要的常量和static方法，代码如下：

```
package qianshou.data;
public class Contant {
   public static int WIDTH = 32;
   public static int HEIGHT = 32;
   public static int LEFT_X = 100;
   public static int LEFT_Y = 50;
   public static int h = LEFT_Y,gap = 0;
   public static synchronized void  moveView(ViewForPerson view,
   boolean boo){
       view.setPersonViewLocation(LEFT_X/4+gap,h);
       if(boo){
         h = h+HEIGHT+10;
       }
   }
}
```

（2）Point.java刻画点的代码如下：

```
package qianshou.data;
import java.util.*;
public class Point{
    int x,y;
    boolean havePerson;
    //散列表存放在当前点上的全部人
    HashMap<String,Person>allPersonsAtPoint;
    public Point(){
       allPersonsAtPoint = new HashMap<String,Person>();
    }
    public void setX(int x){
        this.x=x;
    }
    public void setY(int y){
        this.y=y;
    }
    public boolean isHavePerson(){        //判断当前点上是否有人
        return !allPersonsAtPoint.isEmpty();
    }
    public int getX(){
        return x;
```

```
    }
    public int getY(){
        return y;
    }
    private HashMap<String,Person> addPersonAtPoint(Person person){
        //在点上添加一个person
        if(person!=null) {
           //表示person到达当前点上
           allPersonsAtPoint.put(person.getNumber(),person);
        }
        return allPersonsAtPoint;
    }
    private HashMap<String,Person> getAllPerson(){
      //返回当前点上的全部人
        return allPersonsAtPoint ;
    }
    private HashMap<String,Person> removePerson(Person person){
        if(person!=null)
          allPersonsAtPoint.remove(person.getNumber());
        return allPersonsAtPoint;
    }
   public synchronized HashMap<String,Person>
          handlePersonAtPoint(int type,Person person){
       HashMap<String,Person> mp = null;
       if(type == 0)
          mp = addPersonAtPoint(person);
       else if(type == 1)
          mp = removePerson(person);
       else if(type ==2 )
          mp = getAllPerson();
       return mp;
   }
}
```

（3）Person.java刻画人的代码如下：

```
package qianshou.data;
import java.util.*;
public abstract class Person extends Thread{
    String number;
    Point [][] allCanAtPoint;                    //人可能所在的全部点位置
    Point point;                                 //人当前所在的点位置
    LinkedList<Point> arroudPoints;              //存放和point相邻的点
    ViewForPerson PersonView;                    //人的外观视图
```

```
    boolean successHand;                            //牵手是否成功
    public Person(){
       arroudPoints = new LinkedList<Point>();
    }
    public void setAtPoint(Point p) {
        if(p!=null){
          point = p;
          point.handlePersonAtPoint(0,this);

        }
    }
    public Point getAtPoint() {                     //当前人所在的点
        return point;
    }
    public void setAllCanAtPoint(Point [][] point){
        allCanAtPoint = point;
    }
    public void setPersonView(ViewForPerson PersonView) {
        this.PersonView = PersonView;
        PersonView.setPerson(this);
    }
    public ViewForPerson getPersonView() {
         return PersonView;
    }
    public void setNumber(String s) {
        number = s;
    }
    public String getNumber() {
        return number;
    }
    public abstract boolean move();
}
```

（4）PersonChina.javaPerson的子类的代码如下：

```
package qianshou.data;
import java.util.*;
public class PersonChina extends Person  {
   public boolean move(){
       arroudPoints.clear();
       int m = -1,n=-1;
       boolean successMove = false;
       Point p = getAtPoint();
       for(int i=0;i<allCanAtPoint.length;i++){
```

```
  for(int j=0;j<allCanAtPoint[i].length;j++)
  if(allCanAtPoint[i][j] == p){
      m = i;                                    //找到人当前所处的索引位置
      n = j;
      break;
  }
}
for(int i=0;i<allCanAtPoint.length;i++){
  for(int j=0;j<allCanAtPoint[i].length;j++)
  if(Math.abs(i-m)+Math.abs(j-n)==1){
     arroudPoints.add(allCanAtPoint[i][j]);
  }
}
int size = arroudPoints.size();            //当前周围相邻点的数目size
Random random = new Random();
int k = random.nextInt(size);              //随机获取[0,size)的整数
Point randomPoint = arroudPoints.get(k);

HashMap<String,Person>
allPersonAtPoint =
randomPoint.handlePersonAtPoint(2,this);
                           //正在randomPoint点上的全部人
if(allPersonAtPoint.containsKey(number)){        //假如有牵手的号码

   successHand = true;
   Person lovePerson= allPersonAtPoint.get(number);    //得到牵手之人
   lovePerson.successHand = true;
   ViewForPerson personView =this.getPersonView();
                                           //得到人的视图
   p.handlePersonAtPoint(1,this);          //移出牵手的当前人
   randomPoint.handlePersonAtPoint(1,lovePerson);  //移出被牵手的人
   Contant.moveView(this.getPersonView(),false);
   Contant.gap = Contant.WIDTH-3;
   Contant.moveView(lovePerson.getPersonView(),true);
   Contant.gap = 0;
}
else {
   ViewForPerson personView = getPersonView();//得到人的视图
   personView.setPersonViewLocation(randomPoint.getX(),
   randomPoint.getY());
   p.handlePersonAtPoint(1,this);          //人离开当前点
   this.setAtPoint(randomPoint);           //人到达新的点
   successHand = false;
   successMove = true;
```

```
        }
        return successMove ;
    }
    public void run(){
        while(successHand == false) {    //牵手未成功
            boolean boo = move();        //运动，寻找自己的目标
            try {
               Thread.sleep(410);
            }
            catch(Exception exp){}
        }
    }
}
```

（5）ViewForPerson.java视图接口的代码如下：

```
package qianshou.data;
import javax.swing.JPanel;
public abstract class ViewForPerson extends JPanel {
    public abstract void setPerson(Person person);
    public abstract void setImage(String name);
    public abstract Person getPerson();
    public abstract void setPersonViewLocation(int x,int y);
    public abstract void setPersonViewSize(int w,int h);
}
```

（6）PersonView.java人的视图的代码如下：

```
package qianshou.view;
import qianshou.data.*;
import java.awt.*;
public class PersonView extends ViewForPerson{
   Person person;
   Image image;
   Toolkit tool;
   public PersonView() {
      tool = getToolkit();
   }
   public void setPerson(Person person){
      this.person = person;
   }
   public void setImage(String name){
```

```
        image = tool.getImage(name);
        repaint();
    }
    public Person getPerson() {
        return person;
    }
    public void setPersonViewLocation(int x,int y){
        setLocation(x,y);
    }
    public void setPersonViewSize(int w,int h){
        setSize(w,h);
    }
    public void paintComponent(Graphics g){
        super.paintComponent(g);
        int w=getBounds().width;
        int h=getBounds().height;
        g.drawImage(image,0,0,w,h,this);
        g.setFont(new Font("",Font.BOLD,18));
        g.drawString(person.getNumber(),9,h/2+5);
    }
}
```

（7）LookLoveView.java牵手视图区域的代码如下：

```
package qianshou.view;
import qianshou.data.Point;
import qianshou.data.Person;
import qianshou.data.PersonChina;
import qianshou.data.ViewForPerson;
import qianshou.data.Contant;
import javax.swing.*;
import java.awt.*;
import java.awt.event.*;
import java.util.*;
public class LookLoveView extends JPanel {
    int m = 18,n =31;                              //点阵的行数和列数
    public Point [][] point;                       //存放所有的点
    public Person [] persons ;                     //存放所有的人
    public ViewForPerson [] personView;            //存放人的视图
    String boy = "boy.pgn";
    String girl = "girl.pgn";                      //人的视图上的图像
    int width = Contant.WIDTH;                     //点之间的横向距离
    int height =Contant.HEIGHT;                    //点之间的纵向距离
    int leftX = Contant.LEFT_X;                    //左上角起点偏移坐标
```

```
int leftY = Contant.LEFT_Y;
int [] hao ={1,1,2,2,3,3,4,4,5,5,6,6,7,7,
            8,8,9,9,10,10,11,11,12,12,
            13,13,14,14,15,15};
int personAmount = hao.length;                      //人数
public LookLoveView(){
   setLayout(null);
   setBackground(Color.white);
   initPointXY();                                   //初始化point点
   initPerson();                                    //初始化人
   validate();
   repaint();                                       //绘制区域线条
}
public void start(){
   for(int i = 0; i< persons.length;i++){
      persons[i].start();
   }
}
public void initPointXY(){                          //依据视图设置点的坐标
  //组件坐标系原点是左上角，向右是x轴，向下是y轴
   point = new Point[m][n];
   for(int i=0;i<point.length;i++) {
     for(int j=0;j<point[i].length;j++){
       point[i][j] = new Point();
       point[i][j].setX(j*width+leftX);
       point[i][j].setY(i*height+leftY);
     }
   }
}
public void initPerson(){
   persons = new Person[personAmount];
   for(int i = 0; i< persons.length;i++){
      persons[i] = new PersonChina();
      persons[i].setAllCanAtPoint(point);
      persons[i].setNumber(""+hao[i]);
   }
   initPersonView();                                //初始化人的视图
   initPositon();                                   //初始化人及视图位置
}
public void initPersonView(){                       //初始化人物的视图
   personView = new PersonView[persons.length];
   for(int i = 0; i< personView.length;i++){
      personView[i] = new PersonView();
      if(i%2==0)
```

```
            personView[i].setImage(girl);
        else
            personView[i].setImage(boy);
        persons[i].setPersonView(personView[i]);  //人物的视图外观
        personView[i].setPerson(persons[i]);
    }
}
public void initPositon(){    //初始化人及视图位置
    for(int i = 0; i< persons.length;i++){
        add(personView[i]);    //将人的视图容器添加到容器中
        personView[i].setPersonViewSize(width,height);
    }
    LinkedList<Point> list = new LinkedList<Point>();
    Random random = new Random();
    for(int i=0;i<m;i++) {
        for(int j=0;j<n;j++){
            list.add(point[i][j]);
        }
    }
    for(int i = 0; i< persons.length;i++){
        //随机得到一个[0,list.size())的数
        int index =random.nextInt(list.size());
        //从链表中删除一个节点，并得到该节点
        Point personAtpoint = list.remove(index);
        persons[i].setAtPoint(personAtpoint);
        int x = personAtpoint.getX();
        int y = personAtpoint.getY();
        persons[i].getPersonView().setPersonViewLocation(x,y);
    }
}
public void paintComponent(Graphics g){  //绘制点的视图外观
    super.paintComponent(g);
    Color c = new Color(200,200,200);
    g.setColor(c);
    for(int i = 0;i<point.length;i++){
        for(int j = 0;j<point[i].length;j++)
            g.drawRect(point[i][j].getX(),point[i][j].getY(),
            width, height);
    }
    g.setColor(Color.blue);
    g.setFont(new Font("楷体",Font.PLAIN,20));
    g.drawString("默默想一个号码,看能否牵手"+
    "。随机走动，同号相遇在同一点上，即可牵手",Contant.LEFT_X+10,Contant.
    LEFT_Y-10);
```

```
        g.drawString(
        "软件编写:耿祥义",Contant.LEFT_X+10,Contant.LEFT_Y+m*Contant.
        HEIGHT+25);
    }
}
```

（8）WindowView.java主类的代码如下：

```
package qianshou.view;
import qianshou.data.Point;
import qianshou.data.*;
import java.awt.*;
import javax.swing.JFrame;
import java.util.ArrayList;
public class WindowView extends JFrame{
    LookLoveView loveView;
    public WindowView(){
        loveView = new LookLoveView();
        add(loveView,BorderLayout.CENTER);
        setBounds(5,5,1260,730);
        setDefaultCloseOperation(JFrame.EXIT_ON_CLOSE);
        setVisible(true);
        loveView.start();
    }
    public static void main(String args[]) {
        new WindowView();
    }
}
```

二十四节气

36.1 场景故事

地球有自转和公转。地球相对太阳有一个倾斜角，地球自转轴线和运动平面之间的夹角是66.5°，不是90°（就像人行走时身体不直，此处可以想想地球仪）。地球围绕太阳转动一周，使得太阳在地球上的直射点形成的椭圆轨迹和地球本身的赤道（椭圆）不在一个平面上，二者之间的夹角是23.5°，如图36.1所示。

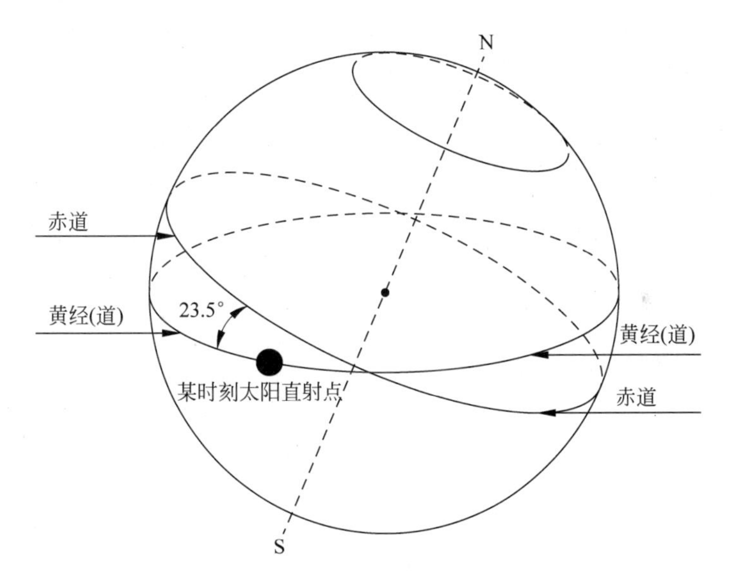

图36.1 赤道、黄道与倾斜角

太阳在地球上的直射点形成的椭圆称作黄经（或黄道）。中国的古人就是根据黄经计算出了二十四节气。因为是根据太阳直射点的位置计算出的二十四节气，因此节气在公历的每年都略有不同，就像《节气歌》唱的那样：每月两节不变更，最多相差一两天。比如，当太阳的直射点是黄经315°时就是立春（直射南纬11°45'），即通常所说的立春时太阳达到黄经315°。立春通常是在2月4日，比如，2019年立春是2月4日11:14:14（北京时间，此时美国是深夜），立秋是8月8日03:12:57。而2018立春是2月4日05:28:25，立秋是8月7日21:30:34。春分时太阳达到黄经0°（直射赤道），秋分时太阳达到黄经315°（直射赤道），冬至时直射南回归线（然后开始北移），夏至时直射北回归线（然后开始南移）。如果把黄经（道）放大，那么可以刚好和地球围绕太阳的公转椭圆轨道吻合，即当太阳的直射点在黄经（道）上某个位置时，地球也恰好运动到公转轨道的相应的位置，比如太阳直射点是黄经315°，地球恰好运动到公转轨道的315°，即太阳直射点到达黄道315°，地球运动到达轨道315°。

36.2 场景故事的目的

1. 侧重点

计时器Timer类、内部类以及绘制基本图形、图像的知识，特别是绘制圆弧的知识。

使用Arc2D.Double类创建圆弧对象：

```
new Arc2D.Double(double x,double y,double w, double h, double start,double extent,int type)
```

圆弧是椭圆的一部分。参数x、y、w、h指定椭圆的位置和大小，参数start和extent的单位都是“度”。参数start、extent表示从start的角度开始沿逆时针或顺时针方向画出extent度的弧。当extent是正值时表示逆时针，否则表示顺时针。比如，起始角度start是0表示3点钟的方位，start的值可以是负值，例如-90度表示6点的方位。其中，最后一个参数type的取值可以是Arc2D.OPEN、Arc2D.CHORD、Arc2D.PIE，分别表示弧是开弧、弓弧或饼弧。

2.涉及的其他知识点

匿名类和Lambda表达式。如果一个接口里只有一个方法，那么可以用Lambda表达式简化使用实现接口的类的类体创建对象的代码，将代码

```
new 接口名() { 接口中方法名(参数 a,参数 b,...){... } }
```

简化为：

```
(参数 a,参数 b,...) ->{... }
```

3.进一步的尝试

增加背景音乐功能。

36.3 程序运行效果与视频讲解

视频讲解

将模拟季节点以及程序需要的其他图像保存在image文件夹中，并让image文件夹和主类Earth位于相同的目录中。主类是Earth。二十四节气程序运行效果如图36.2所示。

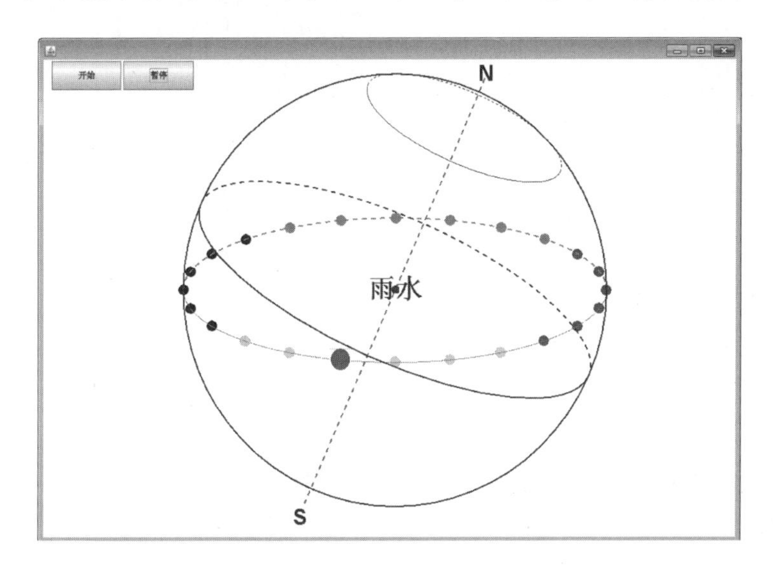

图36.2 二十四节气程序运行效果

36.4 阅读源代码

Earth.java的代码如下：

```
import javax.swing.*;
import java.awt.*;
import java.awt.geom.*;
import java.awt.event.*;
class MyCanvs extends Panel implements ActionListener {
     final int STARTANGLE =225;        //立春，相当于天文学上的315度
     Toolkit tool;
     Image [] imageSeason;             //季节图标
```

```
SeasonLabel [] seasonLabel;       //放置季节图标的组件
Image imageSun;                   //太阳直射在地球上的图像
JButton buttonStart,buttonStop;
SunLabel sun;                     //模拟太阳直射点的图像
Timer time;                       //动画计时器
boolean isStart;
double startAngle = STARTANGLE;  //立春，相当于天文学上的315度
int width = 600;                  //椭圆的宽
int height = 200;                 //椭圆的高
int m =22;
int translationX = 200;           //图形x方向平移量
int translationY = 220;           //图形y方向平移量
Ellipse2D earthPoint;             //地心小圆
Line2D axisLine;                  //地球自转的轴线
Arc2D huangjingArc;               //黄经实线弧
Arc2D huangjingDottedArc;         //黄经虚线弧
Arc2D equatorArc;                 //赤道实线弧
Arc2D equatorDottedArc;           //赤道虚线弧
Ellipse2D earth;                  //地球的圆示意图
Arc2D seasonSunArc;               //太阳直射点形成的椭圆（即黄经）
int sunSize =30;                  //模拟直射点形成的太阳的大小
int seasonSize =15;               //季节点的大小
int seasonAngle[] =
{225,240,255,270,285,300,315,330,345,0,15,30,45,60,75,90,105,120,
135,150,165,180,195,210};
String [] seasonName ={"立春","雨水","惊蛰","春分","清明","谷雨",
                       "立夏","小满","芒种","夏至","小暑","大暑",
                       "立秋","处暑","白露","秋分","寒露","霜降",
                       "立冬","小雪","大雪","冬至","小寒","大寒"};
JLabel nameSeaon;                    //显示当前节气名
MyCanvs() {
  imageSeason = new Image[24];
  seasonLabel = new SeasonLabel[24];
  nameSeaon = new JLabel("",SwingConstants.CENTER);
  nameSeaon.setFont(new Font("宋体",Font.BOLD,38)); //80
  setLayout(null);
  setBackground(Color.white);
  sun = new SunLabel();
  startAngle = STARTANGLE;
  tool=getToolkit();
  for(int i = 0;i<imageSeason.length; i++ )  {
      imageSeason[i] = tool.getImage("image/"+i+".jpg");
      //0.jpg是立春的图标
      seasonLabel[i] = new SeasonLabel(imageSeason[i]);
```

```
        add(seasonLabel[i]);
        seasonLabel[i].setSize(seasonSize,seasonSize);
        seasonLabel[i].setVisible(false);
    }
    imageSun=tool.getImage("image/sun.jpg");
    time = new Timer(7,this);
    buttonStart = new JButton("开始");
    buttonStop = new JButton("暂停");
    add(buttonStart);
    add(buttonStop);
    add(sun);
    add(nameSeaon);
    sun.setVisible(false);
    buttonStart.setBounds(12,2,100,40);
    buttonStop.setBounds(115,2,100,40);
    sun.setSize(sunSize,sunSize);
    nameSeaon.setSize(100,100);                    //300,300
    nameSeaon.setLocation(translationX+width/2-50,
    translationY+height/2-50);
    buttonStop.addActionListener((ActionEvent e)->{ time.stop();});
     //用Lambda表达式
    buttonStart.addActionListener((ActionEvent e)->{ time.start();});
    earthPoint = new Ellipse2D.Double(width/2-6,height/2-6,12,12);
      //地心小圆
    axisLine =
    new Line2D.Double
    (width/2,width/2+height/2+m,width/2,-(width-height)/2-m);
                                                  //地球自转的轴线
    earth=
    new Ellipse2D. Double (0,-(width-height)/2,width,width);
                                                  //地球(圆)
    huangjingArc =
    new Arc2D.Double(0,0,width,height,0,-180,Arc2D.OPEN); //黄经实线弧
    huangjingDottedArc =
    new Arc2D.Double(0,0,width,height ,0,180,Arc2D.OPEN); //黄经虚线弧
    equatorDottedArc =
    new Arc2D.Double(0,0,width,height,0,180,Arc2D.OPEN); //赤道虚线弧
    equatorArc =
    new Arc2D.Double(0,0,width,height,0,-180,Arc2D.OPEN); //赤道实线弧
    seasonSunArc =
    new Arc2D.Double(0,0,width,height,STARTANGLE-1,-360,Arc2D.OPEN);
}
public void actionPerformed(ActionEvent e) {
     Point2D point = seasonSunArc.getStartPoint();
```

```
    int x = (int)point.getX();
    int y = (int)point.getY();
    seasonSunArc.setAngleStart(startAngle);
    for(int i=0;i<=23;i++) {
      if(Math.abs(startAngle-seasonAngle[i]) <= Math.pow(10,-8)) {
        seasonLabel[i].
        setLocation(x+translationX-seasonSize/2,
                    y+translationY-seasonSize/2);
        seasonLabel[i].setVisible(true);
        nameSeaon.setText(seasonName[i]);
      }
    }
    startAngle += 0.05;
    startAngle = startAngle%360;
    sun.setLocation(x+translationX-sunSize/2,y+translationY-sunSize/2);
    sun.setVisible(true);
    sun.repaint();
}
public void paint(Graphics g) {
     super.paint(g);
     paintBasic(g);                              //绘制基本图形图像
}
class SunLabel extends JPanel{                   //刻画太阳直射点的样子
       SunLabel(){
           setOpaque(true);
       }
       public void paint(Graphics g) {
          super.paint(g);
          g.drawImage
          (imageSun,0,0,getBounds().width,getBounds().height,this);
                                                 //平铺绘制图像
       }
}
class SeasonLabel extends JPanel{                //季节小图标
       Image imageSeason;
       SeasonLabel(Image imageSeason){
          this.imageSeason = imageSeason;
          setOpaque(false);
       }
       public void paint(Graphics g) {
          super.paint(g);
          g.drawImage
          (imageSeason,0,0,getBounds().width,getBounds().height,
          this);                                 //平铺绘制图像
```

```
        }
}
public void paintBasic(Graphics g) { sun.repaint();
   buttonStart.repaint();
   buttonStop.repaint();
   Graphics2D g_2d=(Graphics2D)g;
   g_2d.setFont(new Font("",Font.BOLD,30));
   g_2d.drawString("N",618,30);
   g_2d.drawString("S",355,645);
   AffineTransform trans=new AffineTransform();
   trans.translate(translationX,translationY); //trans设置为平移变换
   g_2d.setTransform(trans);
   BasicStroke strokeDotte = new
   BasicStroke(2, 1, 1, 1, new float[]{6,7} ,1);        //设置虚线笔头
   g_2d.setStroke(strokeDotte);
   g_2d.setColor(new Color(181,155,36));
   g_2d.draw(huangjingDottedArc);
   BasicStroke stroke = new BasicStroke(1, 1, 1, 1); //设置实线笔头
   g_2d.setStroke(stroke);
   g_2d.draw(huangjingArc);
   trans.rotate(23.5*Math.PI/180, width/2, height/2);
                                                    //trans设置为旋转变换
   g_2d.setColor(Color.blue);
   g_2d.setTransform(trans);
   strokeDotte = new
   BasicStroke(2, 1, 1, 1, new float[]{6,7} ,1);        //设置虚线笔头
   g_2d.setStroke(strokeDotte);
   g_2d.draw(equatorDottedArc);
   strokeDotte = new
   BasicStroke(2, 1, 1, 1, new float[]{6,7} ,1);        //设置虚线笔头
   g_2d.setStroke(strokeDotte);
   g_2d.setColor(Color.red);
   g_2d.draw(axisLine);
   stroke = new BasicStroke(2, 1, 1, 1);                //设置实线笔头
   g_2d.setStroke(stroke);
   g_2d.setColor(Color.blue);
   g_2d.draw(earth);
   g_2d.draw(equatorArc);
   g_2d.fill(earthPoint);
   //调整绘制北极附近的小椭圆
   trans.translate(150,-195);
   trans.scale(0.5,0.5);
   g_2d.setTransform(trans);
   g_2d.draw(equatorArc);
```

```
        g_2d.setStroke(strokeDotte);
        g_2d.draw(equatorDottedArc);
    }
}
public class Earth{
   public static void main(String args[]) {
      JFrame win = new JFrame();
      win.setSize(1000,700);
      win.add(new MyCanvs());
      win.setDefaultCloseOperation(JFrame.EXIT_ON_CLOSE);
      win.setVisible(true);
   }
}
```

附 录 A

Java核心内容之归纳与概括

A.1 基本语法

1. 标识符

标识符不能是关键字以及true、false和null。标识符由字母、下画线、美元符号和数字组成并且第一个字符不能是数字。

2. 基本数据类型

基本数据类型按级别从低到高包括：byte、short、char、int、long、float、double。当把级别高的值赋给级别低的变量时，必须使用类型转换运算，即显式运算。

3. 数组

数组属于引用型变量。两个相同类型的数组如果具有相同的引用，它们就有完全相同的元素。编译时无须检查数组索引是否越界，但程序运行时，一旦发现数组索引越界就会触发ArrayIndexOutOfBoundsException异常。

4. 算术混合运算的精度

对于byte、short、char、int、long、float和double，当计算表达式值时，按表达式中出现的最高精度的值进行计算。如果表达式中最高精度的值低于int型，则按int精度进行运算。Java没有byte、short型的常量，例如23、100等都是int常量，即按4字节处理。23L、100L是long常量，按8字节处理。允许把不超出byte、short或char型的取值范围、精度为int的常量的算术表达式的值赋给byte、short或char型变量。但是，对于含有变量的表达式，必须进行显式类型转换。

5. 关系表达式与逻辑表达式

关系以及逻辑表达式的值是true或false。逻辑运算符的操作元必须是boolean型数据。

6. 控制语句与循环语句

if-else控制语句以及while、do-while和for循环语句中的条件表达式的值必须是boolean型数据。switch语句中条件表达式的值可以为byte、short、int、char、String或enum（枚举）型，但不可以是long型。switch语句中的“常量值1”至“常量值n”对应的也是byte、short、int、char、String 或enum型，而且要互不相同。

A.2 核心基础

1. 类

类的目的是抽象出一类事物共有的属性和行为，并用一定的语法格式来描述所抽象出的属性和行为，即封装了数据和对数据的操作。类是一种用于创建具体实例（对象）的数据类型。“类体”的内容由两部分构成：一部分是变量的声明，另一部分是方法的定义。对成员变量的操作只能放在方法中，方法使用各种语句对成员变量和方法体中声明的局部变量进行操作。

2. 构造方法

构造方法的名字必须与它所在的类的名字完全相同，而且没有类型。允许一个类中编写若干个构造方法，即构造方法也可以被重载。如果类中没有显式地定义构造方法，则系统会默认该类只有一个构造方法，该默认的构造方法是无参数的，且方法体中没有语句。如果类中定义了一个或多个构造方法，那么Java不提供默认的构造方法。new运算符可以和类的构造方法进行运算，运算过程是：首先为成员变量分配内存，并指定默认值，然后初始化成员变量（类声明成员变量时给定的默认值），接着执行构造方法，最后计算出一个引用值，即new运算符的结果是引用值（通过将该对象的内部地址，即分配给对象的变量地址信息，转换成一个整数得到引用值）。

3. 对象的结构

类是一种用于创建具体实例（对象）的数据类型，类声明的变量称为一个对象变量，简称为对象。对象（变量）负责存放引用，以确保对象可以操作分配给该对象的变量（实体）以及调用类中的方法（体现对象的行为、功能）。两个类型相同的对象，一旦二者的引用相同，二者就具有完全相同的变量（实体）。不要把对象和分配给对象的变量两者混淆（分配给对象的变量仅仅是对象的一部分）。通常，习惯性地把对象归类到“引用型”变量。

4. 组合

如果一个对象a组合了对象b，那么对象a就可以委托对象b调用其方法，即对象a以组合的方式复用对象b的方法。通过组合对象来复用方法也称为“黑盒”复用，因为当前对象只能委托所包含的对象调用其方法，这样一来，当前对象对所包含的对象的方法的细节（算法的细节）是一无所知的。当前对象随时可以更换所包含的对象，即对象与所包含的对象属于弱耦合关系。

5. static关键字

实例变量仅仅是和相应的对象关联的变量，也就是说，不同对象的实例变量互不相同，即分配不同的内存空间，改变其中一个对象的实例变量不会影响其他对象的这个实例变量。对象的实例变量可以通过该对象访问，但不能使用类名访问。类变量（static变量）是与该类创建的所有对象相关联的变量，改变其中一个对象的这个类变量就同时改变了其他对象的这个类变量。因此，类变量不仅可以通过某个对象访问，还可以直接通过类名访问。

实例方法中不仅可以操作实例变量，还可以操作类变量。当对象调用实例方法时，该方法中出现的实例变量就是分配给该对象的实例变量，该方法中出现的类变量也是分配给该对象的变量，只不过这个变量和所有其他对象共享而已。类方法（static方法）不仅可以被对象调用执行，还可以直接通过类名调用。类方法不可以操作实例变量，只能操作类变量。如果一个方法不需要操作类中的任何实例变量，就可以满足程序的需要，进而可以考虑将这样的方法设计为一个类方法。

6. 方法重载

一个类中允许多个方法具有相同的名字，但这些方法的参数必须不同。即参数的个数不同或参数个数相同，但参数列表中对应的某个参数的类型不同。

7. this关键字

this关键字代表某个对象，可以出现在实例方法和构造方法中，但不可以出现在类方法中。this关键字出现在类的构造方法中时，代表使用该构造方法所创建的对象。实例方法只能通过对象来调用，不能用类名来调用，当this关键字出现在实例方法中时，this就代表正在调用该方法的当前对象。

8. 子类的继承性

子类继承父类的成员变量作为自己的一个成员变量，就好像它是在子类中直接声明一样，可以被子类中定义的任何实例方法操作，如果子类中定义的实例方法不能操作父类的某个成员变量，该成员变量就没有被子类继承。

子类继承父类的方法作为子类中的一个方法，就像它是在子类中直接定义的一样，可以被子类中定义的任何实例方法调用。子类不继承父类的构造方法（因为构造方法必须和其类名相同，并且构造方法没有类型。但是，其他方法必须有类型）。子类和父类在同一个包中，子类继承父类中不是private的成员变量作为自己的成员变量，继承父类中不是private的方法作为自己的方法。当子类和父类不在同一个包中时，子类只继承父类中的protected和public访问权限的成员变量作为子类的成员变量，只继承父类中的protected和public访问权限的方法作为子类的方法。

9. 成员变量的隐藏

子类声明的成员变量的名字和从父类继承来的成员变量的名字相同，子类就会隐藏所继承的成员变量。子类对象以及子类定义的方法操作与父类同名的成员变量是指子类重新声明的这个成员变量。子类继承的方法可以操作子类继承或隐藏的成员变量，但不可以操作子类声明的成员变量。

10. 方法的重写

如果子类可以继承父类的某个方法，那么子类就有权利重写这个方法。在子类中定义一个方法，这个方法的类型和父类方法的类型一致或者是父类方法的类型的子类型，并且这个方法的名字、参数个数、参数的类型和父类的方法完全相同。子类如此定义的方法称作子类重写的方法。重写时，不可以降低方法的访问权限，不可以将实例方法重写为static方法，也不可以将static方法重写为实例方法。子类一旦重写了父类的方法，就隐藏了继承的方法，那么子类对象所调用的方法一定是子类重写的方法。重写方法既可以操作继承的成员变量，调用继承的方法，又可以操作子类新声明的成员变量，调用新定义的其他方法，但无法操作被子类隐藏的成员变量和方法。

11. 上转型对象

上转型对象可以访问子类继承或隐藏的成员变量，调用子类继承的方法或子类重写的实例方法。上转型对象操作子类继承的实例方法或子类重写的实例方法，其作用等同于使用子类对象调用这些方法。子类重写了实例方法后，上转型对象调用这个实例方法时，一定是调用了子类重写的实例方法。需要注意的是，子类即使重写了父类的static方法，上转型对象仍然调用的是父类的static方法（从软件设计角度看，子类重写父类的static方法意义不大）。不要将父类创建的对象和子类对象的上转型对象混淆。

12. super关键字

子类一旦隐藏了继承的成员变量，那么子类创建的对象就不再拥有该变量，该变量将归关键字super所拥有；同样，子类一旦隐藏了继承的方法，那么子类创建的对象就不能调用被隐藏的方法，该方法的调用由关键字super负责。因此，如果在子类中想使用被子类隐藏的成员变量或方法，就需要使用关键字super。例如，super.x、super.play()就是访问和调用被子类隐藏的成员变量x和play()方法。

13. 面向抽象编程

当设计某种重要的类时，不是让该类面向具体的类，而是面向抽象类（abstract类），即所设计类中的重要数据是抽象类声明的对象，而不是具体类声明的对象。面向抽象编程的目的是为了应对用户需求的变化，将某个类中经常因需求变化而需要改动的代码从该类中分离出去，即将类中每种可能的变化对应地交给抽象类的一个子类去负责。

14. 接口回调

接口属于引用型变量，接口变量中可以存放实现该接口的类的实例的引用，即存放对象的引用。接口回调是指可以把实现某接口的类的对象的引用赋值给该接口声明的接口变量，那么该接口变量就可以调用该类实现的接口方法。

15. 面向接口编程

当设计某个重要的类时，不让该类面向具体的类，而是面向接口，即所设计类中的重要数据是接口声明的对象，而不是具体类声明的对象。面向接口编程的目的是为了应对用户需求的变化，将某个类中经常因需求变化而需要改动的代码从该类中分离出去，即将类中每种可能的变化对应地交给实现接口的一个类去负责。

16. Class类

Class是java.lang包中的类。Class的实例封装和类有关的信息，即“类型”信息，如该类有哪些构造方法、哪些成员变量、哪些方法等。使用Class的类方法

```
public static Class<?> forName(String className) throws ClassNotFoundException
```

可以返回一个和参数className指定的类相关的Class对象。再让这个Class对象调用

```
public Constructor<?> getDeclaredConstructor() throws SecurityException
```

方法，得到className类的无参数的构造方法（要求className类必须有无参数的构造方法）。然后Constructor<?>对象调用newInstance()方法，返回一个className类的对象。

17. 匿名类与Lambda表达式

直接使用一个类的子类的类体创建对象，称为用一个匿名类（匿名子类）创建对象。直接使用实现接口的类的类体创建对象，称为用一个匿名类（实现接口的匿名类）创建对象。

如果一个接口里只有一个方法，那么可以用Lambda表达式简化匿名类创建对象的代码，也就是将代码（匿名类创建对象）

```
new 接口名() { 接口中方法名(参数a,参数b,...){... } }
```

简化为

```
(参数a,参数b,...)->{... }
```

18. 枚举类型

使用关键字enum定义枚举类型。例如：

```
public enum TrafficLight {
   red,yellow,green;
}
```

声明了一个枚举类型后，就可以用该枚举类型声明一个枚举变量。例如：

```
TrafficLight light;
```

声明了一个枚举变量light。枚举变量light的取值只能是枚举类型中的常量。使用枚举名和“.”运算符可以获得枚举类型中的常量，例如：

```
light = TrafficLight.green;
```

另外，枚举类型中也可以像类一样声明变量以及定义方法。

A.3 应用基础

1. String类

java.lang包中的String类为final类，即String类不可以有子类。String对象封装字符序列和操作字符序列的许多方法。String对象封装的字符序列是不可被修改的（String类不提供这样的方法），String对象属于不可变对象。

2. StringTokenizer类

java.util包中的 StringTokenizer类用分隔标记创建封装“单词”的实例（对象）。其实例常用于获取“单词”。

3. Scanner类

java.util包中的Scanner类用分隔标记创建封装“正则表达式”的实例（对

象）。其实例常用于获取“单词”。

4. Pattern类与Match类

Match类和Pattern类隶属于java.util.regex包。Match类的对象直接检索出和Pattern类的对象匹配的String对象，因此，在某些应用中，Match和Pattern类更容易解决某些检索问题。

5. StringBuffer类

java.lang包中的StringBuffer对象封装字符序列和操作字符序列的许多方法。StringBuffer对象封装的字符序列是可被修改的，StringBuffer类提供了可修改字符序列的许多方法。

6. LocalDate类、LocalDateTime类和LocalTime类

java.time包中的LocalDate类、LocalDateTime类和LocalTime类的对象封装和日期、时间有关的数据。这三个类都是final类，而且不提供修改数据的方法，即这些类的对象的实体不可再发生变化，属于不可变对象。

7. Math类、BigInteger类和Random类

java.lang包中的Math类包含许多用来进行科学计算的类方法，这些方法可以直接用类名调用。java.math包中的BigInteger类提供任意精度的整数运算。使用java.util包中的Random类的实例可以获取随机值。

8. GUI事件处理

GUI（Graphics User Interface）事件处理基于授权模式，即事件源调用方法将某个对象注册为自己的监视器。Java语言使用接口回调技术实现处理事件的过程，在Java中能触发事件的对象都用形如addXXXListener(XXXListener listener)的方法将某个对象注册为自己的监视器，该方法中的参数listener是一个接口，可以引用任何实现了该接口的类所创建的对象，事件源触发事件时，接口listener立刻回调类实现的接口中的某个方法。从方法绑定角度来看，Java运行系统要求监视器必须绑定某些方法来处理事件，这就需要用接口来达到此目的。将事件的某种处理绑定

到对应的接口，即绑定到接口中的方法。换言之，事件源触发事件后，监视器准确知道去调用哪个方法（自动去调用）。“监视器”和“事件源”应该保持一种松耦合关系，即尽量让事件源所在的类和监视器构成组合关系，事件源触发事件发生后，系统知道某个方法会被执行，但无须关心到底是哪个对象去调用了这个方法，因为任何实现接口的类的实例都可以调用这个方法来处理事件。

9. MVC结构

模型-视图-控制器（Model-View-Controller）是一种通过3个不同部分构造一个软件或组件的理想办法。模型（Model）用于存储数据的对象；视图（View）为模型提供数据显示的对象；控制器（Controller）用于处理用户的交互操作，对于用户的操作做出响应，让模型和视图进行必要的交互，即通过视图修改、获取模型中的数据，当模型中的数据变化时，让视图更新显示。

A.4 专项应用

1. 输入、输出流

java.io包（I/O流库）提供大量的流类。抽象类InputStream（字节输入流）或抽象类Reader（字符输入流）的子类称作输入流，抽象类OutputStream（字节输出流）或抽象类Writer（字符输出流）的子类称作输出流。输入流使用read()方法读入源中的数据，输出流使用write()方法把数据写入目的地。

2. 线程

线程是比进程更小的执行单位。一个进程在其执行过程中可以产生多个线程，形成多条执行线索，每条线索即每个线程都有它自身的产生、存在和消亡的过程。每个Java应用程序都有一个默认的主线程。当JVM加载代码、发现main()方法之后，就会启动一个线程，这个线程称为主线程（main线程），该线程负责执行main()方法。那么，在main()方法的执行时再创建的线程，就称为程序中的其他线程。JVM保证让Java应用程序中的多个线程都有机会使用CPU资源，即让多个线程轮流执行。

新建的线程在它的一个完整的生命周期中通常经历如下的4种状态：新建状

态（NEW）、可运行状态（RUNNABLE）、中断状态（BLOCKED，WAITING，TIMED_WAITING）和死亡状态（TERMINATED）。一个线程完成了它的全部工作之后，即执行完run()方法，该线程进入死亡（TERMINATED）状态。

可以用Thread的子类创建线程，当线程调用start()方法，被JVM转入运行状态后，所执行的就是子类重写的run()方法。可以用Thread类创建线程，并指定线程的目标对象，创建目标对象的类必须实现Runnable接口，当线程调用start()方法，被JVM转入运行状态后，目标对象就执行类实现的Runnable接口中的run()方法。

3. 网络编程

Java提供专门直接用于网络编程的URL、Socket、InetAddress和DatagramSocket类。

URL类是java.net包中的一个重要的类，URL的实例封装着一个统一资源定位符（Uniform Resource Locator，URL），使用URL创建对象的应用程序称为客户端程序。一个URL对象封装着一个具体的资源的引用，表明客户要访问这个URL中的资源，客户利用URL对象可以获取URL中的资源。InetAddress类对象封装着Internet主机地址的域名和IP地址。当两个程序需要使用TCP通信时，可以使用Socket类。当两个程序需要使用UDP通信时，可以使用DatagramSocket类。

4. 集合框架

集合框架涉及的类均在java.util包中。

（1）LinkedList<E>实现了泛型接口List<E>，LinkedList<E>泛型类的对象以链表结构存储数据，习惯上称LinkedList类创建的对象为链表。使用LinkedList<E>泛型类声明创建链表时，必须要指定E的具体类型，然后链表就可以使用add(E obj)方法向链表中依次增加节点。

（2）ArrayList<E>实现了泛型接口List<E>，ArrayList<E>泛型类的对象采用顺序结构来存储数据，习惯上称ArrayList类创建的对象为数组表。使用ArrayList<E>泛型类声明创数组表时，必须要指定E的具体类型，然后数组表就可以使用add(E obj)方法向数组表中依次增加节点。

（3）Stack<E>实现了泛型接口List<E>，Stack<E>泛型类的对象采用栈式结构存储数据（先进后出），习惯上称Stack类创建的对象为堆栈。使用Stack<E>泛型类声明创建堆栈时，必须要指定E的具体类型，然后堆栈就可以使用void push(E item)实现压栈操作，使用public E pop()实现弹栈操作。

（4）HashMap<K,V>实现了泛型接口Map<K,V>，HashMap<K,V>类的对象采用散列表这种数据结构存储数据，习惯上称HashMap<K,V>对象为散列映射。使用HashMap<K,V>泛型类声明创建散列映射时，必须要指定K和V的具体类型。散列映射用于存储“键/值”对，允许把任何数量的“键/值”对存储在一起。键不可以发生逻辑冲突，即不能让两个数据项使用相同的键，如果出现两个数据项对应相同的键，则此前散列映射中的“键/值”对将被替换。

（5）TreeSet<E>类实现Set<E>接口，TreeSet<E>类创建的对象称为树集，使用TreeSet<E>泛型类声明创建树集时，必须要指定E的具体类型，然后树集就可以使用add(E obj)方法向树集中依次增加节点。树集采用树结构存储数据，当一个树集中的对象是实现Comparable接口的类创建的对象时，节点就按照对象的大小关系顺序排列。树节点中的数据会按存放的数据的“大小”顺序一层一层地依次排列。

（6）HashSet<E>泛型类实现了泛型接口Set<E>，习惯上称HashSet<E>对象为集合。使用HashSet<E>泛型类声明创建集合时，必须要指定E的具体类型，然后集合就可以使用add(E obj)方法向集合中添加元素。HashSet<E>泛型类的对象在数据组织上类似数学上的集合，可以进行“交”“并”“差”等运算。集合不允许有相同的元素，也就是说，如果b已经是集合中的元素，那么再执行set.add(b)操作是无效的（默认情况下，如果两个元素的引用相同，则它们属于相同的元素）。

使用HashSet集合可能经常需要处理特殊情况：“可能两个元素的引用不同，程序不希望集合中同时有这两个元素”。例如，要从电话号码中选择两位尾号分别是95、85、75和65的电话一部。有许多电话的号码最后两位是95，却是不同的电话（不同的元素），如果按照HashSet集合的默认验证条件，这些电话都可以添加到集合中。处理办法如下所述。

① 重写hashCode()方法。Phone类重写从Object类继承的int hashCode()方法（返回一个int常量即可），让所有电话的“散列值”相同，即废除“散列值”的作用。这样做的原因是，当HashSet集合添加新元素时，首先验证集合中是否有和其“散列值”相同的元素（HashSet集合使用元素的hashCode()方法返回的值作为元素的“散列值”），如果没有就直接将该元素添加到集合中，否则再用当前元素的equals()方法（从Object类继承的方法）检查集合中是否有和当前集合相等的元素（默认情况下，如果两个元素的引用相同就属于相同的元素）。

② 重写equals()方法。Phone类可以重写equals()方法，重新规定Phone对象相等的条件。

（7）Collections类提供了将List中的数据重新排序的

```
public static <T extends Comparable<? super T>> void sort(List<T>
list)
```

方法。只要List中节点的对象实现了Comparable<T>泛型接口中的public int compareTo(T t)方法即可。

例如，对于

```
LinkedList<Rect> rectList = new LinkedList<Rect>();
```

如果Rect实现了Comparable<T>泛型接口，例如：

```
class Rect implements Comparable<Rect>{
    public int width;
    public int height;
    public int compareTo(Rect rect){
      return width = rect.width
    }
}
```

那么Collections.sort(rectList)就会将rectList节点中的对象按int compareTo(Rect rect)方法规定的大小关系（按矩形的宽度比大小）重新排序（从小到大）。

Collections类还提供了将List中的数据重新排序的方法如下：

```
public static <T> void sort(List<T> list, Comparator<? super T> c)
```

此方法不要求List中节点中的对象实现Comparable<T>泛型接口中的public int compareTo(T t)方法，而是在Collections调用此方法的过程中，动态实现Comparator<T>泛型接口中的int compare(T o1,T o2)方法。

例如，对于

```
LinkedList<Rect> rectList = new LinkedList<Rect>();
class Rect {
   public int width;
   public int height;
}
```

那么

```
Collections.sort(rectList,(Rect o1,Rect o2)->{return o1.height-o2.height;})
```

就会将rectList节点中的对象按int compare(Retc o1,Rect o2)方法规定的大小关系（按矩形高度比大小）重新排序（这里用到了Lambda表达式）。

Collections类还提供了将链表中的数据重新随机排列的类方法以及旋转链表中数据的类方法如下：

public static void shuffle(List<E> list) 将list中的数据按洗牌算法重新随机排列。

static void rotate(List<E> list, int distance) 旋转链表中的数据。例如，假设list的数据依次为10、20、30、40、50，那么调用Collections.rotate(list,1)之后，list的数据依次为50、10、20、30、40。当方法的参数distance取正值时，向右转动list中的数据，取负值时向左转动list中的数据。

public static void reverse(List<E> list) 翻转list中的数据。假设list索引处的数据依次为1、2、3，那么调用Collections. reverse (list)之后，list的数据依次为3、2、1。

5．绘制图形与播放音频

java.awt包中的Component类有一个方法public void paint(Graphics g)，程序可以在其子类中重写这个方法。当程序运行时，java运行环境会用Graphicd2D（Graphics的一个子类）将参数g实例化，对象g就可以在重写paint()方法的组件内绘制图形、图像等。组件都是矩形形状，组件本身有一个默认的坐标系，组件左上角的坐标值是(0,0)。如果一个组件的宽是200，高是80，那么，该坐标系中，x坐标的最大值是200，y坐标的最大值是80。

java.awt.geom包提供的Graphics2D拥有强大的二维图形处理能力。Graphics2D是Graphics类的子类，它把直线、矩形、椭圆、圆弧等作为一个对象来绘制，也就是说，如果想用一个Graphics2D对象来画一个圆的话，就必须先创建一个圆的对象。Graphics2D对象分别使用draw和fill方法来绘制和填充一个图形。

javax.sound.sampled包中的Clip对象是一个守护线程，帮助播放音频文件。步骤如下所述。

（1）得到Clip对象。代码如下：

```
Clip clip = AudioSystem.getClip();
```

（2）得到音频流。代码如下：

```
AudioInputStream stream= AudioSystem.getAudioInputStream(new File
("音频文件"));
```

（3）clip打开音频流。代码如下：

```
clip.open(stream);
```

（4）播放或暂停音频流。代码如下：

```
clip.start();
clip.stop();
```

（5）关闭音频流。代码如下：

```
clip.close();
```

6.JDBC

Java提供了专门用于操作数据库的API，即Java数据库连接（Java Data Base Connectivity，JDBC）。JDBC操作不同的数据库时，仅仅是连接方式上的差异而已，使用JDBC的应用程序一旦和数据库建立连接，就可以使用JDBC提供的API操作数据库，如图A-1所示。

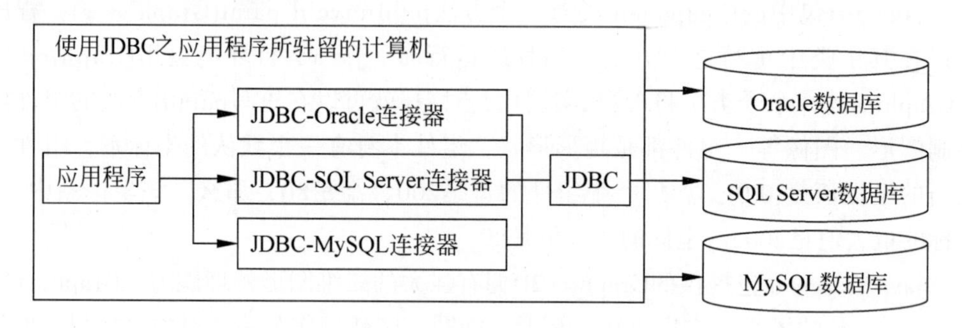

图A-1　使用JDBC操作数据库